Chang Wook Ahn

Advances in Evolutionary Algorithms

Studies in Computational Intelligence, Volume 18

Editor-in-chief
Prof. Janusz Kacprzyk
Systems Research Institute
Polish Academy of Sciences
ul. Newelska 6
01-447 Warsaw
Poland
E-mail: kacprzyk@ibspan.waw.pl

Further volumes of this series
can be found on our homepage:
springer.com

Vol. 4. Saman K. Halgamuge, Lipo Wang
(Eds.)
*Classification and Clustering for Knowledge
Discovery*, 2005
ISBN 3-540-26073-0

Vol. 5. Da Ruan, Guoqing Chen, Etienne E.
Kerre, Geert Wets (Eds.)
Intelligent Data Mining, 2005
ISBN 3-540-26256-3

Vol. 6. Tsau Young Lin, Setsuo Ohsuga,
Churn-Jung Liau, Xiaohua Hu, Shusaku
Tsumoto (Eds.)
*Foundations of Data Mining and Knowledge
Discovery*, 2005
ISBN 3-540-26257-1

Vol. 7. Bruno Apolloni, Ashish Ghosh, Ferda
Alpaslan, Lakhmi C. Jain, Srikanta Patnaik
(Eds.)
Machine Learning and Robot Perception,
2005
ISBN 3-540-26549-X

Vol. 8. Srikanta Patnaik, Lakhmi C. Jain,
Spyros G. Tzafestas, Germano Resconi,
Amit Konar (Eds.)
Innovations in Robot Mobility and Control,
2006
ISBN 3-540-26892-8

Vol. 9. Tsau Young Lin, Setsuo Ohsuga,
Churn-Jung Liau, Xiaohua Hu (Eds.)
*Foundations and Novel Approaches in Data
Mining*, 2005
ISBN 3-540-28315-3

Vol. 10. Andrzej P. Wierzbicki, Yoshiteru
Nakamori
Creative Space, 2005
ISBN 3-540-28458-3

Vol. 11. Antoni Ligęza
*Logical Foundations for Rule-Based
Systems*, 2006
ISBN 3-540-29117-2

Vol. 13. Nadia Nedjah, Ajith Abraham,
Luiza de Macedo Mourelle (Eds.)
Genetic Systems Programming, 2006
ISBN 3-540-29849-5

Vol. 14. Spiros Sirmakessis (Ed.)
Adaptive and Personalized Semantic Web,
2006
ISBN 3-540-30605-6

Vol. 15. Lei Zhi Chen, Sing Kiong Nguang,
Xiao Dong Chen
*Modelling and Optimization of
Biotechnological Processes*, 2006
ISBN 3-540-30634-X

Vol. 16. Yaochu Jin (Ed.)
Multi-Objective Machine Learning, 2006
ISBN 3-540-30676-5

Vol. 17. Te-Ming Huang, Vojislav Kecman,
Ivica Kopriva
*Kernel Based Algorithms for Mining Huge
Data Sets*, 2006
ISBN 3-540-31681-7

Vol. 18. Chang Wook Ahn
Advances in Evolutionary Algorithms, 2006
ISBN 3-540-31758-9

Chang Wook Ahn

Advances in Evolutionary Algorithms

Theory, Design and Practice

 Springer

Dr. Chang Wook Ahn
Samsung Advanced Institute
of Technology (SAIT)
14-1 Nongseo-Dong
Kiheung-Gu, Gyeonggi-Do
Republic of Korea, 446-712
E-mail: cwan@evolution.re.kr

ISSN print edition: 1860-949X
ISSN electronic edition: 1860-9503
ISBN 978-3-642-06860-7   e-ISBN 978-3-540-31759-3

Springer is a part of Springer Science+Business Media
springer.com
© Springer-Verlag Berlin Heidelberg 2006
Softcover reprint of the hardcover 1st edition 2006

To my parents

Preface

The goal of this book is to develop efficient optimization algorithms to solve diverse real-world problems of graded difficulty. Genetic and evolutionary mechanisms have been deployed for reaching the goal.

This book has made five significant contributions in the realm of genetic and evolutionary computation (GEC).

Practical guidelines for developing genetic algorithms (GAs) to solve real-world problems have been proposed. This fills a long standing gap between theory and practice of GAs. A practical population-sizing model for computing solutions with desired quality has also been developed. The model needs no statistical information about the problems. It has duly been validated by computer simulation experiments.

The suggested design-guidelines have been followed in developing a GA for solving the shortest path (SP) routing problem. Experimental studies validate the effectiveness of the guidelines. Further, the population-sizing model passes the feasibility test for this application. It appears to be applicable to a wide class of problems.

Elitist compact genetic algorithms (cGAs) have been developed under the framework of simple estimation of distribution algorithms (EDAs). They can deal with memory- and time-constrained problems. In addition, they do not require any prior knowledge about the problems. The design approach enables a typical cGA to overcome selection noise. This is achieved by persisting with the current best solution until, hopefully a better solution is found. A higher quality of solutions and a higher rate of convergence are attained in this way for most of the test problems. The hidden connection between EDAs and evolutionary strategies (ESs) has been made explicit. An analytical justification of this relationship is followed by its empirical verification. Further, a speedup model that quantifies convergence improvement has also been developed. Experimental evidence has been supplied to support the claims.

The real-coded Bayesian optimization algorithm (rBOA) has been proposed under the general framework of advanced EDAs. Many difficult problems – especially those that can be decomposed into subproblems of bounded difficulty – can be solved quickly, accurately, and reliably with rBOA. It can automatically discover unsuspected problem regularities and effectively exploit this knowledge to perform robust and scalable search. This is achieved by constructing the Bayesian factorization graph using finite mixture models. All the relevant substructures are extracted from the graph. Independent fitting of each substructure by mixture distributions is then followed by drawing new solutions by independent subproblem-wise sampling. An analytical model of rBOA scalability in the context of problems of bounded difficulty has also been investigated. The criterion that has been adopted for the purpose is the number of fitness function evaluations until convergence to the optimum. It has been shown that the rBOA finds the optimal solution with a sub-quadratic scale-up behavior with regard to the size of the problem. Empirical support for the conclusion has also been provided. Further, the rBOA is found to be comparable (or even better) to other advanced EDAs when faced with nondecomposable problems.

Finally, a competent multiobjective EDA (MEDA) has also been developed by extending the (single-objective) rBOA. The multiobjective rBOA (MrBOA) is able to automatically discover and effectively exploit implicit regularities in multiobjective optimization problems (MOPs). A selection method has been proposed for preserving diversity. This is done by assigning fitness to individuals by domination rank with some penalty imposed on sharing and crowding of individuals. It must be noted that the solution quality is not compromised in the process. It is experimentally demonstrated that MrBOA outperforms other state-of-the-art multiobjective GEAs (MGEAs) for decomposable as well as nondecomposable MOPs.

It is thought that this work will have a major impact on future genetic and evolutionary computation (GEC) research. Our ardent hope is that it will play a decisive role in bringing about a paradigm shift in computational optimization research.

December 2005   *Chang Wook Ahn*

Acknowledgements

There are several people who helped write this book. I would like to convey my gratitude to them.

Foremost, I would like to thank my parents for their absolute and continuous love, dedication and trust that were the prime source in finishing up this work. I am also thankful to the rest of my family, my sisters and my late grandmother, for their endless love and support.

I would like to acknowledge Prof. R. S. Ramakrishna gratefully. I could not have taken this delight without his guidance with valuable advice and a deep affection. I am sincerely thankful to Prof. David E. Goldberg for the invaluable comments and suggestions on this work. Especially, he allowed me the great opportunity of working with him and other members of the Illinois Genetic Algorithms Laboratory (IlliGAL). I would also like to express my gratitude to Prof. Hyoung Woo Lee and Prof. Chung Gu Kang who led me towards real-academic world and improved my research ability. Also, I would like to thank all the professors of the department of information and communications in the Gwangju Institute of Science and Technology (GIST).

I am sincerely grateful to a number of friends and colleagues whom I met during my visit to the IlliGAL, Dr. Martin Butz, Dr. Jian-Hung Chen, Dr. Ying-Ping Chen, Nazan Khan, Dr. Xavier Llorà, Dr. Kei Onishi, Gerulf Pederson, Kumara Sastry, Abhishek Sinha, Tian-Li Yu, for their kindness and help. I am also thankful to Dr. Martin Pelikan and Dr. Jiri Ocenasek for their interests and opinions.

With a view to improving the quality of this book, any comments and suggestions are deeply appreciated. *cwan@evolution.re.kr* is available for correspondence.

Abbreviations

BB	Building Block
BIC	Bayesian Information Criterion
BMOA	Bayesian Mutiobjective Optimization Algorithm
BOA	Bayesian Optimization Algorithm
cGA	Compact Genetic Algorithm
EA	Evolutionary Algorithm
ecGA	Extended Compact Genetic Algorithm
EDA	Estimation of Distribution Algorithm
EGNA	Estimation of Gaussian Networks Algorithm
ES	Evolutionary Strategy
FDA	Factorized Distribution Algorithm
FDA_c	Continuous Factorized Distribution Algorithm
GA	Genetic Algorithm
GEA	Genetic and Evolutionary Algorithm
GEC	Genetic and Evolutionary Computation
hBOA	Hierarchical Bayesian Optimization Algorithm
IDEA	Iterative Density-estimation Evolutionary Algorithm
ISG	Ising Spin-Glasses
m(h)BOA	Multiobjective (Hierarchical) Bayesian Optimization Algorithm
MBOA	Mixed Bayesian Optimization Algorithm
mDP	Minimal Deceptive Problem
MDP-I	Multiobjective Deceptive Problem I
MDP-II	Multiobjective Deceptive Problem II
MEDA	Multiobjective Estimation of Distribution Algorithm
MGEA	Multiobjective Genetic and Evolutionary Algorithm
mIDEA	Mixed Iterative Density-estimation Evolutionary Algorithm
MIDEA	Multiobjective Iterative Density-estimation Evolutionary Algorithm
MNSP	Multiobjective Nonlinear, Symmetric Problem
MOGA	Multi-Objective Genetic Algorithm

MOP	Multiobjective Optimization Problem
MrBOA	Multiobjective Real-coded Bayesian Optimization Algorithm
ne-cGA	Nonpersistent Elitist Compact Genetic Algorithm
NPGA	Niched Pareto Genetic Algorithm
NSGA	Nondominated Sorting Genetic Algorithm
NSGA-II	Nondominated Sorting Genetic Algorithm II
PBBC	Probabilistic Building-Block Crossover
pdf	Probability Density Function
pe-cGA	Persistent Elitist Compact Genetic Algorithm
PMBGA	Probabilistic Model Building Genetic Algorithm
PV	Probability Vector
rBOA	Real-coded Bayesian Optimization Algorithm
RDGA	Rank-Density-based Genetic Algorithm
RDP	Real-valued Deceptive Problem
RLA	Randomized Leader Algorithm
RMOP	Real-valued Multiobjective Optimization Problem
RNSP	Real-valued Nonlinear, Symmetric Problem
sGA	Simple Genetic Algorithm
SNR	Signal to Noise Ratio
SP	Shortest Path
SPEA	Strength Pareto Evolutionary Algorithm
SPEA-II	Strength Pareto Evolutionary Algorithm II
UMDA	Univariate Marginal Distribution Algorithm
$UMDA_c$	Continuous Univariate Marginal Distribution Algorithm

Contents

1

Introduction

Every *real-world* problem from economic to scientific and engineering fields is
ultimately confronted with a common task, viz., *optimization* [1, 3, 20, 38, 89].
An optimization problem can be defined by specifying the set of all feasible
candidates and a measure for evaluating their worth [89]. The goal is to find
the best solution(s). In the design of aerofoils, for instance, the parameters that
define the geometry of the aerofoil are optimized to achieve the desired surface
pressure distribution. In the design of a satellite antenna, the antenna pattern
is optimized to maximize the mainbeam gain while minimizing the sidelobe
gain. In robot trajectory planning, the position, orientation, velocity, and
acceleration that specify robot trajectory are optimized for feasible obstacle
free motion.

Intense research activity over the years has resulted in many optimization algorithms. They are, however, still limited in their reach. In this regard,
there is growing interest in the design of adaptive optimization techniques.
It makes an attempt to discover and exploit invisible (problem) patterns in
solving various real-world problems in an efficient and scalable manner. This
is similar to *black-box* optimization [20, 89]. In black-box optimization, there
is no prior information about the relation between the performance measure
and the semantics of the solutions. However, the knowledge can be gathered by sampling new candidate solutions and assessing their suitability (i.e.,
quality). Some well known techniques in this regard include random search,
hill climbing, and so forth. A well structured traversal of the search space
incorporates state-of-the-art computing technologies such as computational
intelligence. Genetic and evolutionary algorithms (GEAs) belong to a class of
the advanced black-box optimization algorithms.

GEAs evolve a population of promising solutions by following a two-
operator mechanism – *selection* and *variation*. They emulate some natural
processes. The population approach eliminates noise in evaluating solution
quality. It allows simultaneous search of multiple basins of attraction. The selection operator nudges the search toward superior solutions, whereas the variation operators promote wider exploration. Recombination (or crossover) and

Chang Wook Ahn: *Advances in Evolutionary Algorithms: Theory, Design and Practice*, Studies
in Computational Intelligence (SCI) **18**, 1–5 (2006)
www.springerlink.com

mutation are the commonly used variation operators [11, 32, 38, 48]. Recombination promotes purposeful search by combining superior partial solutions; while mutation overcomes local traps by slightly perturbing current solutions. The trust in these algorithms may be misplaced in that they turn out to be more and more expensive as the number of parameters (of the problem) increases. The central theme of this book is related to these issues.

1.1 Motivation

GEAs have an enviable success record in solving real-world problems in diverse areas [3, 23, 38, 48, 52, 84]. In some sense, they offer a panacea to practitioners in a wide range of disciplines. Significant progress has been registered in the theory and design of competent GEAs [9, 40, 41, 45, 73]. They can efficiently deal with very hard optimization problems. Despite considerable theoretical achievements, GEA practitioners often discern a gap between *theory* and *practice*. This is acutely felt when they try to design algorithms for real-world problems. There has been little or no effort to bridge this gap, however.

A new GEA paradigm has received attention of late. This is the *estimation of distribution algorithms* (EDAs), also known as *probabilistic model building genetic algorithms* (PMBGAs) [63,64,89,90]. EDAs are good at automatic discovery and exploitation of problem regularities. They combine unique features of GEAs (viz., genetic inheritance and survival of the fittest) with advanced computing methods of machine learning and (graphical) probabilistic modeling. Based on the intricacy of the probabilistic model, EDAs are roughly divided into two categories – *simple* and *advanced*. The simple approach incurs no computational cost for discovering and exploiting problem regularities, but it is extravagant on solution quality evaluations. The advanced approach works in just the opposite way.

The simple approach is quite promising for some real-world applications such as unicast or multicast routing, call admission control, resource allocation, and so forth. In these problems, a matter of primary importance is to find acceptable solution(s) as quickly as possible (i.e, real-time requirement). One can offer to be liberal on the number of inexpensive solution quality computations. Meanwhile, the advanced approach is apt for a class of real-world problems such as DNA array analysis, space-station structure design, etc. This is because optimality of the computed solution(s) is of primary importance here and high computational cost is a necessary "evil".

The simple approach cannot be directly applied to real-world problems involving real-time and limited-memory constraints. Even though these problems are relatively easy to solve, there are some difficulties related to deception and interactions between decision variables. It is possible to devise a variant that lies somewhere in between simple and advanced schemes by restricting the complexity of the probabilistic model [14, 26, 87]. However, the computational cost for providing prior information on problem regularities can be unacceptably high. Moreover, its overall complexity leaves much to be desired. Consequently, new simple EDAs must be devised for effectively coping with such issues. Some results [15,52,86] reported in this context still require excessive computational resources.

In general, many important real-world problems have some complicated structures. A representative example is a pattern of interactions between decision variables. Without knowing the inherent features, it is quite hard to find optimal solution(s). This has motivated researchers to design *competent* algorithms. Several advanced (discrete) EDAs for solving difficult real-world problems are known. They decompose a problem into several subproblems of bounded difficulty and then intermix their desirable features [44,61,76,88,89]. Their effectiveness has been well supported by tests on artificial as well as actual real-world problems. The discrete EDAs have led to similar work on continuous (i.e., real-valued) problems [20,63,82]. However, the attempts have not been very successful.

Many real-world problems have multiple irreconcilable and often competing objectives. These problems are known as multiobjective optimization problems (MOPs) . The goal of multiobjective optimization is to find a complete set of solutions (i.e., *Pareto-optimal* set) such that no other solutions in the search space are better than them with respect to all the considered objectives. Many multiobjective genetic and evolutionary algorithms (MGEAs) have been reported [19,29,37,38,68,122]. They choose promising candidates that facilitate convergence to global Pareto-optimal set while maintaining uniform spread of the candidates . In other words, there has been little or no effort to develop competent MGEAs that efficiently identify, propagate, and intermix important partial solutions of the problem. The sequence of procedures is a critical factor in devising successful MGEAs (as in single-objective GEAs).

1.2 Objectives

In the light of the above discussion, the following five primary objectives have been set for this book.

1. Establish useful guidelines for designing practical GEAs as a class of optimization algorithms.
2. Design a genetic algorithm (GA) for solving the shortest path (SP) routing problem following the suggested practical guidelines.
3. Design a class of simple but efficient optimization algorithms under the framework of simple EDAs.
4. Develop a competent optimization algorithm in the context of advanced EDAs for solving problems in the continuous domain.
5. Extend the competent EDA (in the fourth objective) with a view to dealing with the multiobjective optimization.

The first objective will play a critical role in filling the gap between theory and practice in designing practical GEAs for dealing with a broad class of real-world applications. The second objective will demonstrate the practical utility of the suggested design road map. The third objective will offer a useful tool to significantly enhance the exploratory power in time-constrained and memory-limited applications. The fourth objective will lead to a class of promising (scalable) procedures that are capable of solving hard problems in the continuous domain. The problems are assumed to be decomposable into subproblems of bounded difficulty. The last objective will open an important track for MGEA research that relies on discovering and utilizing problem regularities of MOPs.

The objectives appear to have real importance because they are intended to make GEAs highly promising in dealing with *simple* to *hard*, *time-* to *quality-constrained*, and *single-* to *multi-objective* real-world (optimization) problems in a wide range of disciplines.

1.3 Outline

An outline of this book is given as follows.

Chapter 2 introduces principles of a basic class of GEAs (i.e., GAs) and design-decomposition theory that is critical to successful design. The chapter suggests methodologies for designing GAs for solving real-world problems. A practical population-sizing model is also presented. It facilitates computation of solutions with the desired quality without demanding any prior statistical information about the problems.

Chapter 3 develops a GA for solving the SP routing problem along the lines of design guidelines presented in Chap. 2. The aim of this development is to demonstrate the utility of the guidelines. The population-sizing model is also validated in the context of the routing problem.

Chapter 4 presents a class of elitism-based compact genetic algorithms (cGAs) as simple but efficient EDAs. The design objective is to compensate for inherent defects (of compact-type GAs) connected with lack of memory through elitism. This enables the algorithms to efficiently and speedily solve time- and memory-constrained problems without any overheads on discovering and utilizing problem regularities. Also, some theoretical aspects of the proposed algorithms are investigated.

Chapter 5 describes real-coded Bayesian optimization algorithm (rBOA) as a competent advanced EDA in the continuous domain. It tries to bring the power of existing (discrete) Bayesian optimization algorithm (BOA) to bear upon the area of real-valued (i.e., numerical) optimization. Thus, it can deal with a hard problem by decomposing it into tractable subproblems and then combining the computed partial solutions of the subproblems. Scalability of the rBOA is also analyzed and verified.

Chapter 6 presents multiobjective real-coded Bayesian optimization algorithm (MrBOA). It is an extended version of the proposed rBOA that incorporates the features of multiobjective optimization. It can automatically discover regularities of multiobjective optimization problems and then utilize the knowledge for exploring the search space on the basis of the decomposition principle. This chapter also describes a new selection method that goads current solutions to converge to the set of nondominated solutions while maintaining an appreciable (solution) spread.

Finally, Chap. 7 summarizes and concludes the book. Some directions for future work are also suggested.

2

Practical Genetic Algorithms

Over the last decade, genetic algorithms (GAs) have been successfully applied to problems in business, engineering, and science. This is a consequence of a noteworthy progress in their theory, design and development [3, 11, 25, 38, 41, 48]. In spite of considerable work on various aspects of GAs, practitioners often face hurdles in confronting real-world problems due to inadequate design guidelines. They are often at a loss to come up with proper parameter values for want of relevant theoretical basis. Unavailability of problem dependent information complicates the issue in practice.

This chapter is an attempt to bridge this gap. The chapter also develops a practical population-sizing model. The model helps compute solutions with desired quality, and – this is important – it does so without the aid of any statistical information about the problems.

The chapter is organized as follows. Section 2.1 briefly introduces the principal ideas behind GAs and GA design theory based on the decomposition principle. Section 2.2 suggests some (useful) practical design guidelines. In Sect. 2.3, the population-sizing model is developed and verified. The chapter concludes with a summary in Sect. 2.4.

2.1 Genetic Algorithms: Simple to Competent

This section provides background information on simple genetic algorithms (sGAs). A brief introduction to design decomposition that is necessary to design *competent* GAs is also presented.

2.1.1 Overview of Genetic Algorithms

Genetic algorithms (GAs) are stochastic, population-based search and optimization algorithms inspired by the process of natural selection and genetics [11, 38, 48, 53]. A major characteristic of GAs is that they work with a

Chang Wook Ahn: *Advances in Evolutionary Algorithms: Theory, Design and Practice*, Studies in Computational Intelligence (SCI) **18**, 7–22 (2006)
www.springerlink.com
© Springer-Verlag Berlin Heidelberg 2006

Simple Genetic Algorithm

STEP 1. INITIALIZATION
   Generate initial population $\mathcal{P}$ at random or with prior knowledge
STEP 2. FITNESS EVALUATION
   Evaluate the fitness for all individuals in $\mathcal{P}$
STEP 3. SELECTION
   Select a set of promising candidates $\mathcal{S}$ from $\mathcal{P}$
STEP 4. CROSSOVER
   Apply crossover to the mating pool $\mathcal{S}$ for generating a set of offspring $\mathcal{O}$
STEP 5. MUTATION
   Apply mutation to the offspring set $\mathcal{O}$ for obtaining its perturbed set $\mathcal{O}'$
STEP 6. REPLACEMENT
   Replace the current population $\mathcal{P}$ with the set of offspring $\mathcal{O}'$
STEP 7. TERMINATION
   If the termination criteria are not met, go to STEP 2

Fig. 2.1. Pseudo-code for sGA.

population, unlike other classical approaches which operate on a single solution at a time. Hence, they can explore different regions of the solution space (i.e., search space) concurrently, thereby exhibiting enhanced performance. The pseudo-code of sGAs is shown in Fig. 2.1.

Essential Components

GAs are powerful search mechanisms: traverse the solution space in search of optimal solutions. GAs encode the decision variables (or input parameters) of the underlying problem into (solution) strings. Each string, called *individual* or *chromosome*, represents a candidate solution. Characters of the string are called *genes*. The position and the value in the string of a gene are called *locus* and *allele*, respectively. There are two encoding classes: *genotype* and *phenotype*. The former denotes the codings of the variables and the latter represents the variables themselves.

A fitness function is needed for differentiating between good and bad solutions. Unlike classical optimization techniques, the fitness function of GAs may be presented in a mathematical terms, or as a complex computer simulation, or even in terms of subjective human evaluation. *Fitness* generates a differential signal in accordance with which GAs guide the evolution of solutions to the problem [25].

The initial population is created at random or with prior knowledge about the problem. The individuals are evaluated to measure the quality of candidate solutions with a fitness function. In order to generate or evolve the offspring (i.e., new solutions), genetic operators are applied to the current population. The genetic operators are: *selection* (or *reproduction*), *crossover* (or *recombination*), and *mutation*.

Genetic Operators

Selection chooses the individuals with higher fitness as parents of the next generation. In other words, selection operator is intended to improve average quality of the population by giving superior individuals a better chance to get copied into the next generation. There is a *selection pressure* that characterizes the selection schemes. It is defined as the ratio of the probability of selection of the best individual in the population to that of an average individual [9,73]. There are two basic types of selection scheme in common usage: *proportionate* and *ordinal* selection. Proportionate selection picks out individuals based on their fitness values relative to the fitness of the other individuals in the population. Examples of such a selection type include roulette-wheel selection [38,53], stochastic remainder selection [16], and stochastic universal selection [12]. Ordinal selection selects individuals based not upon their fitness, but upon their rank within the population. The individual are ranked according to their fitness values. Tournament selection [21], (μ, λ) selection [105], linear ranking selection [12], and truncation selection [73] are included in the ordinal selection type.

Crossover exchanges and combines partial solutions from two or more parental individuals according to a crossover probability, p_c, in order to create offspring. That is, the crossover operator exploits the current solutions with a view to finding better ones. Two popular crossover operators, from among many variants, are presented: *one-point* and *uniform* crossover. One-point crossover [38,53] randomly chooses a crossover point (i.e., crossing site) in the two individuals and then exchanges all the genes behind the crossover point (see Fig. 2.2(a)). Uniform crossover [111] exchanges each gene with probability 0.5 (see Fig. 2.2(b)), hence achieving the maximum allele-wise mixing rate.

Mutation acts by altering a small percentage of genes in the list of individuals to slightly perturbs the recombined solutions. One classical mutation operator is *bit-wise* mutation [38,53] in which each gene whose allele is binary is complemented with a mutation probability p_m. For instance, a binary individual A = 1 1 1 1 1 1 might become A' = 1 1 0 1 1 1 when the third gene is chosen (randomly) for mutation. In general, the mutation probability is taken to be low.

By striking a balance between exploitation (of selection) and exploration (of crossover and mutation), GAs can effectively search the solution space.

2.1.2 Design-Decomposition Theory

In the study on *innovation intuition* (of GAs) by Goldberg [40,41], the combined effect of selection and mutation (these GAs are called *selectomutative*

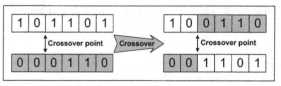

(a) One-point crossover.

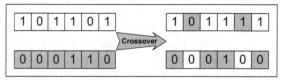

(b) Uniform crossover.

Fig. 2.2. Example of two-parent crossover operators.

GAs) and that of selection and crossover (these GAs are referred to as *selectorecombinative* GAs) has been likened to *continual improvement* and *cross-fertilizing* types of innovation, respectively. On the basis of innovation intuition, a design-decomposition theory has been proposed for developing *competent* (selectorecombinative) GAs, which are a class of GAs that solve hard problems quickly, accurately, and reliably [41]. The design decomposition consists of seven steps briefly described below.

1. **Know what GAs process – Building blocks (BBs):** Competent GAs must decompose the problem into subproblems implicitly (virtually), process them independently (either in a serial or parallel manner), and combine subsolutions to form better solutions or the global optima. The superior subsolutions are identified as *building blocks* (BBs).
2. **Know the BB challenges – BB-wise difficult problems:** Competent GAs should efficiently solve problems of bounded BB difficulty through BB processing. Those problems are known as *decomposable* problems that include a wide range of practical problems.
3. **Ensure an adequate supply of raw BBs:** To successfully solve a problem, all the (necessary) raw BBs must be supplied in the initial population. Although decision-making, mixing, and sampling mainly govern the population-sizing in the evolving population, it would be extremely difficult to maintain the growth of the BBs if one is faced with paucity of BBs.
4. **Ensure increased market share for superior BBs:** The growth of superior BBs in the evolving population is clearly of central importance to ensure a GA success. Thus, competent GAs must give good BBs a higher chance of survival. Note that this issue is closely related to the supply of raw BBs (Step 3), decision making (Step 6), and mixing BBs (Step 7).

5. **Know BB takeover and convergence time:** Although it is necessary to grow the market share for superior BBs, an adequate growth rate is essential. This is because too fast a growth rate often results in premature convergence while too slow a growth rate retards the convergence speed.
6. **Make decisions well among competing BBs:** As increasing the population reduces the noise in decision making, the population size should be large enough to make statistically correct decisions among competing BBs.
7. **Mix BBs well:** Competent GAs should effectively intermix and reassemble superior BBs in order to create promising solutions.

The design-decomposition theory provides valuable guidelines on designing competent GAs. Moreover, it can also be used for investigating the principal mechanisms of GAs and developing theoretical models for predicting the scalability of GAs [25].

2.2 Practical Design Guidelines

Despite the GA design theory – the design decomposition that plays an important role in developing competent GAs, practitioners may still face hurdles due to certain practical issues. This problem is addresses in this section.

There are six issues that lead to practical GA design. These are described below.

1. **Representation:** This issue is primarily related to the encoding scheme. Individuals are represented by binary codes, real-valued (i.e., floating-point) codes, and program code. Moreover, the length of individuals may be constant or variable. In general, it is hard to find an encoding method that transforms a problem so as to reduce or preserve the difficulty of the problem. Hence, the encoding method that has identical genotype and phenotype (of the decision variables) is advisable. Although fixed-length individuals are generally desirable, their variability is not a critical factor provided their design is easy.
2. **Initialization:** In general, there are two issues to be considered for population initialization of GAs: the initial population size and the procedure to initialize the population. At first, the initial population size connected to the supply of raw BBs (in the design-decomposition theory) is crucial for efficiency of GAs in terms of both optimality and complexity. A detailed investigation can be found in Sect. 2.3.4. Secondly, there are two ways to generate the initial population: *random* and *heuristic* initialization. If no prior information on the problem is available, random initialization is the natural choice; otherwise, heuristic initialization is favored. Although the mean fitness of the heuristic initialization is already high so that it may help the GAs to find solutions faster, it may just explore a small part of the solution space and never find global optimal solutions because of lack

of diversity in the population [56]. In the heuristic case, thus, a portion of the population can still be generated randomly to ensure some diversity in the population. It is noted that the random initialization is generally desirable for stability and simplicity of GAs even when a valuable piece of information is available.

3. **Fitness function:** The fitness function interprets the individual in terms of physical representation and evaluates its *fitness* based on desired traits (in the solution). But, the fitness function must accurately measure the quality of the individuals in the population. The definition of the fitness function, therefore, is very crucial. It is suggested that the fitness function fully reflect the physical objective of the problem.

4. **Genetic operators:** The genetic operators must be carefully designed as they directly affect the performance of GAs.

   a) **Selection:** Selection focuses on the exploration of promising regions in the solution space. As proportionate selection is very sensitive to the selection pressure, a scaling function is employed for redistributing the fitness range of the population. The selection pressure of the ordinal selection is independent of the fitness distribution, and is based solely based on the relative ranking of the population although it may also suffer from high selection pressure [9, 73]. In general, the ordinal selection is preferable. Among the selection schemes (in the ordinal selection), tournament selection without replacement is perceived to be effective in achieving low (selection) noise [40]. Recall that tournament selection without replacement works by means of choosing nonoverlapping random sets of s individuals (i.e., tournament size of s) from the population and then selecting the best individual from each set to serve as a parent for the next generation. Typically, the tournament size s is 2 (viz., pairwise tournament), and it would adjust the selection pressure: the selection pressure increases as the tournament size s becomes larger [45, 73]. In this regard, pairwise tournament selection without replacement is advisable.

   b) **Crossover:** Crossover is the primary operator that increases the exploratory power of GAs. In order to successfully achieve the cross-fertilizing type of innovation, crossover operator must ideally intermix good subsolutions without any disruption of the partitions (i.e., BBs). For example, uniform crossover is very promising in the absence of any inter-gene linkage while building-block crossover is better otherwise. Here, building-block crossover uniformly shuffles the genes on the basis of entire partitions (i.e., subsolutions). In practice, uniform crossover is pessimistic as most of real-world problems have the decision variables that are closely interacted each other. Moreover, building-block crossover may also be undesirable because the capability of learning linkage is an essential prerequisite of the operator. Instead of pursuing the maximum BB-wise mixing in the population, it can be also efficient to increase the population size and employ a simple crossover that has a low probability of disrupting the BBs found so far. Therefore, it is recommended that building-block crossover is suitable if the evaluation of fitness function requires a high computational cost; otherwise, one- or two-point crossover is desirable. Naturally, the crossover probability must be relatively high.

c) **Mutation:** Mutation is the secondary operator of GAs to explore a solution space. In other words, a local search is performed in the case of altering nonsalient genes or getting away from local optima is possible when the salient genes are changed. To carry out the continual improvement type of innovation, as in nature, the probability of applying mutation must be very low. Hence, the suggestion with respect of mutation is that any type of mutation designed is applicable as long as its probability is quite small. Moreover, it is possible to get rid of mutation when the design of mutation operator is complicated.

5. **Treating infeasible individuals:** In case that a problem has some constraints, crossover or mutation may often generate infeasible individuals that violate the constraints. There are two strategies to deal with infeasible individuals: one is to impose a penalty and the other is to repair them [56]. A classical method employs penalty functions. It must be noted that the penalty function is critical to ensure quick convergence and high quality of solution. But it is not easy to come up with an appropriate penalty function. Moreover, this technique may sacrifice some feasible individuals as well because the infeasible individuals might continue to be reproduced. On the other hand, the repair method is applied extensively. But it is not always simple to cure infeasible individuals. Hence, the repair strategy is always advisable unless developing a repair function is an arduous task or the designed function is computationally too expensive by far.

6. **Population size:** A problem that arises with GAs is to properly estimate the values of parameters. Most of the parameters can be determined by the transcendental cognition of practitioners so as to attain good performance. However, it is not easy to estimate the population size that guarantees an optimal solution quickly enough. Thus, the population size has generally been perceived as the most important factor. A recent study has developed a refined population-sizing model by integrating the requirements of the BB supply and decision making [45]. It provides an accurate bound on determining an adequate population size that guarantees a solution with desired quality for (selectorecombinative) GAs. However, it requires stochastic information such as the variance of fitness (i.e., noise) and the expected difference value of fitness (i.e., signal) between the best and second-best BBs, which may not be available in many practical problems. With this in view, the practical population-sizing model is suggested in the next section.

2.3 Practical Population-Sizing Model

The question as to how to choose an adequate population size for a particular domain is difficult and has puzzled practitioners for a long time [31, 39, 40, 45, 53, 69]. If the population size is too small, it is not likely that the GAs will find solutions of high quality. However, if the population size is too large, the GAs will unnecessarily waste processing time leading to unacceptably slow convergence. In this section, the practical population-sizing model that ensure a specified quality of solution in investigated by employing the gambler's ruin problem that was considered first by Harik *et al.* [45].

2.3.1 Review of Population-Sizing Models

Holland [53] studied the k-armed bandit problem as a theoretical motivation for GAs. Macready and Wolpert [69] showed a mathematical flaw in Holland's analysis and provided an analytically simple bandit model that is directly applicable to optimization theory.

De-Jong [31] proposed a population-sizing model based on the signal as well as noise characteristics of the k-armed bandit problem. Although the result explicitly exhibited the role of signal-to-noise ratio in estimating population size, the result was unverified and ignored [40, 45].

Goldberg and Rudnick [39] developed the first population-sizing model based on the variance of fitness. Goldberg *et al.* [40] enhanced the model as a conservative bound on the quality of GAs. The population-sizing model permits accurate statistical decision making among competing building blocks. The population-sizing relation conservatively bounds the actual accuracy of GA convergence as long as all major sources of noise (i.e., collateral noise) are considered in the sizing calculation.

Harik *et al.* [45] also develop a population-sizing model by exploiting similarity between the classical random walk problem – the gambler's ruin problem in particular and the selection mechanism of GAs for determining an adequate population size that guarantees a solution of the desired (target) quality. Using test problems that ranged from the simple to the very difficult, the accuracy of the model was verified. The (linear) ranking selection was tacitly assumed because the decision model[1] in [40] is quite appropriate under this selection scheme. It was also assumed implicitly that mutation is not a dominant operator (i.e., crossover-intensive) because it always disrupts BBs. In order to use his results, however, several domain-dependent variables (involved in his decision model) must be known such as the signal that is defined by the fitness difference between the best and second best BBs, the collateral noise that is defined by the root mean square (rms) fitness variance of the BB that is being considered, and the number of BBs in a string. Furthermore, signal-to-noise ratio (SNR), the most important piece of information in Harik's model is

---

[1] It leads directly to the results in [45]

not usually known in practice. His population-sizing model, therefore, is not suitable for applying to the real-world problems.

2.3.2 Harik's Decision Model

The following results follow from Harik's model of selection [45]. Assume that individuals consist of m non-overlapping (i.e., separable) and uniformly scaled BBs of size k. Consider a competition between an individual i_1 with optimal BB H_1 with mean fitness $\bar{f}_{H_1}$ and fitness variance $\sigma_{H_1}^2$, and an individual i_2 with the second best BB H_2 with mean fitness $\bar{f}_{H_2}$ and fitness variance $\sigma_{H_2}^2$. The probability of deciding correctly between these two individuals is the same as the probability that the fitness of i_1 (f_1) is higher than the fitness of i_2 (f_2): the probability that $(f_1 - f_2) > 0$.

The distance between the mean fitness of individual with H_1 ($\bar{f}_{H_1}$) and the mean fitness of individual with H_2 ($\bar{f}_{H_2}$) is denoted by d (i.e., signal). Assuming that the fitness is an additive function of the fitness contributions of all the BBs, f_1 and f_2 are normally distributed (by the *central limit theorem*). Since the fitness distributions of f_1 and f_2 are both normal, the distribution of $(f_1 - f_2)$ is also normal.

The distribution of $(f_1 - f_2)$ is given by [40, 45]

$$(f_1 - f_2) \sim \mathcal{N}(\bar{f}_{H_1} - \bar{f}_{H_2}, \sigma_{H_1}^2 + \sigma_{H_2}^2). \tag{2.1}$$

Substituting d for $(\bar{f}_{H_1} - \bar{f}_{H_2})$ in the above equation, and normalizing, the probability p of making the correct decision on a single trial for the domains where BBs m (that are not competing directly) are independent and equally scaled (i.i.d) is given by [45]

$$p = \Phi\left(\frac{d}{\sqrt{\sigma_{H_1}^2 + \sigma_{H_2}^2}}\right) = \Phi\left(\frac{d}{\sqrt{2m'}\sigma_{bb}}\right) \tag{2.2}$$

where Φ is the cumulative distribution of a normal distribution with zero mean and unit variance, σ_{bb} is the average rms BB standard deviation. Also, $m' = m - 1$ is the total number of collateral noise sources that are not competing directly. The total collateral noise coming from m' is $m'\sigma_{bb}$.

2.3.3 Practical Decision Model

In general, standard deviation can be thought of as the probabilistic "width" or "spread" of distribution of a random variable. Hence, σ_{bb} (i.e., the standard deviation of BBs) indicates the "statistical length" or "spread" of fitness values of BBs from their average fitness value; indeed, the factor $2\sigma_{bb}$ represents the total average range of fitness changes of all the BBs.

Let $\mathcal{X}$ be the average number of competing BBs. Since the signal d is defined as the fitness difference between the best and the second best BBs, from a statistical point of view, the best BB has a fitness value that is the sum of the average and the standard deviation of BBs' fitness, while the second best BB's fitness is the value of subtracting $\{2\sigma_{bb}/(\mathcal{X}-1)\}$ from the best value. This is because it may be assumed that all the competing BBs are ordered and they are distributed uniformly from the best to the worst fitness values. The key point is that the signal d must be small compared with the standard deviation σ_{bb} of BBs, and the interval value between inter-rank BBs is $\{2\sigma_{bb}/(\mathcal{X}-1)\}$. The first assumption is valid when ordinal selection (e.g., pairwise tournament selection without replacement) is employed in real-world problems because the probability that different individuals have the same quality of solution is nearly zero. The second assumption follows from statistical considerations. Moreover, the signal has a small value in practice because there exist suboptimal solutions whose quality is quite comparable with that of the global optimum. Therefore, the signal d can be represented as

$$
\begin{aligned}
d &= \left(\{\bar{f}_{bb} + \sigma_{bb}\} - \left\{ \bar{f}_{bb} + \sigma_{bb} - \frac{2\sigma_{bb}}{\mathcal{X}-1} \right\} \right) \\
&= \left(\{\bar{f}_{bb} + \sigma_{bb}\} - \left\{ \bar{f}_{bb} + \left(1 - \frac{2}{\chi^k-1}\right)\sigma_{bb} \right\} \right) \\
&= \frac{2}{\chi^k-1}\sigma_{bb}
\end{aligned}
\tag{2.3}
$$

where $\bar{f}_{bb}$ is the mean fitness of BBs, χ is the average cardinality of the alphabet, and k is the average order (i.e., size) of BBs. Therefore, Eq. (2.2) can be rewritten as

$$
p = \Phi\left(\frac{2}{\sqrt{2m'(\chi^k-1)}} \right).
\tag{2.4}
$$

As a special case, assume that the cardinality of the alphabet is 2 (i.e., $\chi = 2$) and the size of BB is 1 (i.e., $k = 1$) in a uniformly scaled linear problem, viz., one-max problem. Similar problems were investigated in [45]. It finds the signal d to be 1.0 and the BB variance σ_{bb}^2 to be 0.25, so that the probability p of making the correct decision on a single trial is given by

$$
p = \Phi\left(\frac{d}{\sqrt{2m'}\sigma_{bb}} \right) = \Phi\left(\frac{2}{\sqrt{2m'}} \right).
\tag{2.5}
$$

On the other hand, Eq. (2.4) for the same problem can be rewritten as

$$
p = \Phi\left(\frac{2}{\sqrt{2m'(\chi^k-1)}} \right) = \Phi\left(\frac{2}{\sqrt{2m'}} \right).
\tag{2.6}
$$

This is, of course, the same as Eq. (2.5). It is surprising that the probability of making the correct decision can be obtained by only knowing the average number of BBs of length $m = m' + 1$. No knowledge of signal and noise is required.

2.3.4 Practical Population-Sizing Model

GA succeeds when all the N members of the population in the BBs of interest are correct. From a well-known result from the literature of gambler's ruin problem[2], it follows that the probability $P_{bb}(x_0)$ that the GA eventually succeeds when there are x_0 initial correct BBs is [45]

$$P_{bb}(x_0) = \frac{1 - \left(\frac{q}{p}\right)^{x_0}}{1 - \left(\frac{q}{p}\right)^{N}} \tag{2.7}$$

where $q = 1 - p$ is the probability of losing a copy of the BB in a particular competition.

Since Eq. (2.7) is a conditional probability given that the GA starts with x_0 correct BBs, the probability that the GA succeeds can be found by

$$P_{bb} = \sum_{i=0}^{N} P_{bb}(i) \cdot P(i)$$

$$= \sum_{i=0}^{N} \left[\frac{1 - \left(\frac{q}{p}\right)^{i}}{1 - \left(\frac{q}{p}\right)^{N}}\right] \cdot \binom{N}{i} \left(\frac{1}{\chi^k}\right)^{i} \left(1 - \frac{1}{\chi^k}\right)^{N-i} \tag{2.8}$$

where $P(i)$ represents the probability that the GA starts with i correct BBs. The probability $P(i)$ is connected to the BB supply in the initial population. Using binomial expansion, it can be arranged as follows:

$$P_{bb} = \frac{1 - \left(1 - \frac{1 - \left(\frac{q}{p}\right)}{\chi^k}\right)^{N}}{1 - \left(\frac{q}{p}\right)^{N}} = \frac{1 - \left(1 - \frac{2p-1}{\chi^k p}\right)^{N}}{1 - \left(\frac{1-p}{p}\right)^{N}}. \tag{2.9}$$

The remaining derivation refers to [45]. The P_{bb} may be approximated as

$$P_{bb} \approx 1 - \left(1 - \frac{2p-1}{\chi^k p}\right)^{N}. \tag{2.10}$$

Hence we get

$$N = \frac{\ln(\alpha)}{\ln\left(1 - \frac{2p-1}{\chi^k p}\right)} \tag{2.11}$$

where $\alpha = 1 - P_{bb}$ is the probability of GA failure. Physically, the probability α of GA failure represents the fact that the GA converges to one of the local optimal solutions.

---

[2] It is equivalent to one-dimensional random walk model.

Since $(2p - 1)/(\chi^k p)$ tends to be a small number, $\ln(1 - (2p - 1)/(\chi^k p))$ can be approximated by $-(2p - 1)/(\chi^k p)$. Thus, Eq. (2.11) can be rewritten as follows:

$$N = -\chi^k \ln(\alpha) \frac{p}{2p - 1}. \tag{2.12}$$

An approximate value of p, using the first two terms of the power series expansion of the normal distribution is given by [45]

$$p = \frac{1}{2} + \frac{1}{\sqrt{2\pi}} z. \tag{2.13}$$

From Eq. (2.4), z is found to be $2/(\sqrt{2m'}(\chi^k - 1))$. Thus, a fairly general, practical population-sizing model can be written as follows:

$$N = -\frac{\chi^k}{2} \ln(\alpha) \left(z^{-1} \sqrt{\frac{\pi}{2}} + 1 \right)$$

$$= -\frac{\chi^k}{2} \ln(\alpha) \left(\frac{\chi^k - 1}{2} \sqrt{\pi m'} + 1 \right). \tag{2.14}$$

When the number of genes in the BBs (viz., average order) becomes large, the probability of disrupting the BBs is increased; thus, the population size may be increased to reach a particular quality of solution. This is the reason why a higher probability of disrupting the BBs drives the probability of making the correct decision on a single trial p towards smaller values so that the population size N must be increased for achieving the same GA failure probability α. This can be inferred from Eq. (2.12). However, we can observe an interesting consequence from the experiments of [45]: the population size necessary for obtaining a desired quality of solution is not overly affected by the order of BBs, even if it is considerably large (e.g., $k \geq 4$). Thus, it is as if the population size is not strongly affected by one- or two-point crossover.

Although the mutation operation may disrupt the BBs and retard convergence of BBs, it eventually ensures a better quality of solution by introducing new chromosomes (maintaining the diversity of the population) that help the GA avoid local convergence. Thus, the population will not be increased by the mutation. In other words, the ultimate population size for a solution of desired quality may not be increased by these operations because the minor harmful effects of the crossover are offset by the beneficial effects of mutation.

Equation (2.14) is applicable only to crossover-intensive GA with little or no mutation. There are some approaches in evolutionary computation, and some problems (notably neural networks), that employ mutation as the dominant operator. In these approaches, Eq. (2.14) is not really useful for determining a population size that ensures a solution of desired quality. The evolutionary checkers player coevolved with a fully connected feed forward neural network with an input layer, two hidden layers, and an output node [23] provides a good example in this regard. It needs a tiny population of only 30 to evolve thousands of weights of the neural network.

It must be noted that Eq. (2.14) does not require any knowledge of signal and noise which may not be available in advance in most practical problems. Instead, the model approximates such stochastic information for all the selection mechanisms. Of course, the approximation may induce some discrepancy that depends on the selection mechanism. It can, however, be concluded that the model provides an upper bound[3] for ordinal selection. Also, note that the population-sizing model can be applied to variable-length individuals by employing the average values in the order of BBs and the number of BBs.

2.3.5 Experimental Verification

The practical population-sizing model is verified with test problems of varying difficulty. The test problems include the classical one-max problem and deceptive problems. In all the experiments, pairwise tournament selection without replacement is employed as a typical ordinal selection. A different type of crossover is chosen according to the order of the BBs of each problem. The crossover is applied with probability 1.0 and the mutation probability is set to zero, because the population-sizing model has been developed for the crossover-intensive GA – the only source of diversity is the initial random population. Moreover, the population-sizing model obtained by applying Harik's decision model to Eq. (2.12) is chosen as a reference for investigating how accurately the proposed approach approximates the problem dependent information (i.e, signal d and BB variance σ_{bb}^2). All the results were averaged over 100 independent runs of a simple (generational) GA.

One-Max Problem

We consider the one-max problem that is the most popular test function in research on GAs due to its simplicity. The fitness of an individual is measured as the number of bits set to one. This is a very easy problem for GAs because there is no isolation, deception, and interdependence (of genes) [22, 45]. Since the order of the BBs is one, any crossover does not disrupt them. Thus, uniform crossover with exchange probability 0.5 is employed for achieving the maximum (BB-wise) mixing rate.

Figure 2.3 depicts the results of the population-sizing model on a 100-bit one-max problem. It is seen that the population the experimental results are in agreement with the theory, especially as the population size N increases. Moreover, the practical population-sizing model is perfectly matched with Harik's model because their probabilities of correct decision are equivalent (as explained in Sect. 2.3.3).

---

[3] It means an overestimation of population size to obtain a target quality

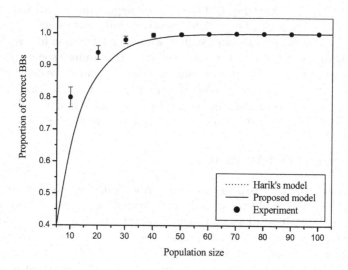

Fig. 2.3. Verification of the population-sizing model for an one-max problem.

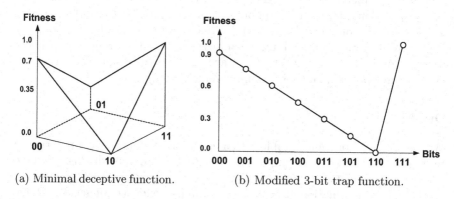

(a) Minimal deceptive function.   (b) Modified 3-bit trap function.

Fig. 2.4. Basis functions of deceptive problems.

Deceptive Problems

Two types of deceptive problem are also considered. The first deceptive problem is a minimal deceptive problem (mDP) that is formed by concatenating twenty copies of the minimal deceptive function [38] shown in Fig. 2.4(a). The second deceptive problem is a fully deceptive problem composed of twenty copies of the modified 3-bit trap function depicted in Fig. 2.4(b). The purpose of the modification is to fulfill the assumptions (described in Sect. 2.3.3) for the target problem. In the deceptive problems, one-point crossover is used for avoiding the excessive disruption of BBs [45].

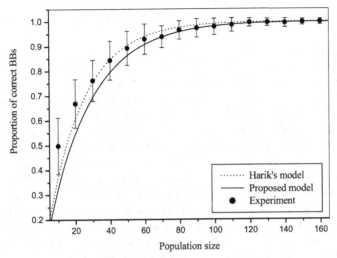

(a) Results for a minimal deceptive problem.

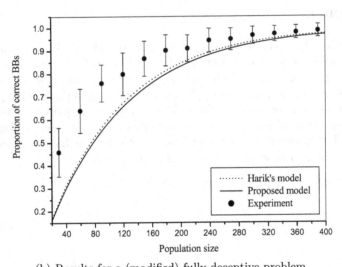

(b) Results for a (modified) fully deceptive problem.

Fig. 2.5. Verification of the population-sizing model for deceptive problems.

The results for deceptive problems are shown in Fig. 2.5. It is also observed that the analytical model is consistent with the experimental results even for higher population size. Moreover, the close agreement between the practical population-sizing model and Harik's model implies that the proposed decision model can accurately approximate the actual SNR without any statistical information about the signal and variance of BBs.

Discussion

From the results of Figs. 2.3 and 2.5, the practical population-sizing model can accurately estimate the actual size of the population with a desired quality of solution without any SNR information. Thus, the results clearly validate the practical population-sizing model.It may be noted that the model would hold true when the fitness difference between the best and second best BBs (i.e., the signal d) is relatively small and all the competing BBs are evenly distributed over the fitness range. However, there is no concern about applying the model because most real-world problems are generally characteristic of satisfying such conditions. Although the population size is overestimated for the optimality below 0.9, such qualities are not regarded as feasible areas in practice. In other words, the model plays a role in providing an upper bound (of population size) with regard to the actual performance.

2.4 Summary

This chapter has sketched a bird's-eye view of GAs. It has also presented the design-decomposition theory that lays guidelines for designing competent GAs. Design of practical GAs for solving real-world problems was the main focus all along. Further, this chapter has also investigated a practical population-sizing model that comes in handy in determining an adequate population size for finding a desired solution without requiring statistical information such as the signal or variance of competing BBs. Its effectiveness has also been tested: the model is in close agreement with experimental results.

Real-World Application: Routing Problem

This chapter presents a genetic algorithmic approach to shortest path (SP) routing problem. Variable-length chromosomes (i.e., strings) and their genes (i.e., parameters) have been used for encoding the problem. The crossover operation exchanges partial chromosomes (i.e., partial routes) at positionally independent crossing sites and the mutation operation introduces new partial chromosomes into the population. The proposed algorithm can cure all the infeasible chromosomes with a simple repair function. Crossover and mutation together provide a search capability that results in improved quality of solution and enhanced rate of convergence.

The chapter is organized as follows. Section 3.1 provides the motivation for considering as powerful tools for dealing with routing problems. A brief survey of GA-based approaches is given in Sect. 3.2. The proposed GA for the SP routing problem is described in Sect. 3.3. In Sect. 3.4, the proposed algorithm and several extant algorithms are applied to diverse networks exhibiting arbitrary link cost, network size, and topology. A comparative study of the results follows. The section also verifies the accuracy of the population-sizing model (developed in Sect. 2.3) in the context of real-world applications. The chapter concludes with a summary in Sect. 3.5.

3.1 Motivation

In multihop networks, such as the Internet and the Mobile *Ad-hoc* Networks, routing is one of the most important issues that has a significant impact on the network's performance [8, 110]. An ideal routing algorithm should strive to find an optimum path for packet transmission within a specified time so as to satisfy the quality of service [1, 8]. There are several search algorithms for the shortest path (SP) problem: the breadth-first search algorithm, the Dijkstra algorithm and the Bellman-Ford algorithm, to name a few [110]. Since these algorithms can solve SP problems in polynomial time, they will be effective in fixed infrastructure wireless or wired networks. But, they exhibit

Chang Wook Ahn: *Advances in Evolutionary Algorithms: Theory, Design and Practice*, Studies in Computational Intelligence (SCI) **18**, 23–43 (2006)
www.springerlink.com   © Springer-Verlag Berlin Heidelberg 2006

unacceptably high computational complexity for real-time communications involving rapidly changing network topologies [1,85]. This is explained below.

We consider mobile *ad hoc* networks as target systems because they represent new generation wireless networks. Since all the nodes cooperatively maintain network connectivity without the aid of any fixed infrastructure networks, dynamic changes in network topology are possible. An optimal (viz., shortest) path has to be computed within a very short time (i.e., a few μs) in order to support time-constrained services such as voice-, video- and teleconferencing [1,8]. The indicated algorithms do not satisfy this (real-time) requirement.

In most of the current packet-switching networks, some form of SP computation is employed by routing algorithms in the network layer [8]. Specifically, the network links are weighted, the weights reflecting the link transmission capacity, the congestion of networks and the estimated transmission status such as the queueing delay of head-of-line (HOL) packet or the link failure. The SP routing problem can be formulated as one of finding a minimal cost path that contains the designated source and destination nodes. In other words, the SP routing problem involves a classical combinatorial optimization problem arising in many design and planning contexts [8,67]. Since neural networks (NNs) [1,8,85] and genetic algorithms (GAs) (and other evolutionary algorithms) [57,67,79,108,120] have been known to offer solutions to such complicated problems, they have also found application in several practical fields. The downside is that NNs and GAs may not be as promising in real-time applications over mobile *ad hoc* networks because they generally involve a large number of iterations. However, hardware implementations (e.g., field-programmable gate arrays (FPGA) chip) of NNs or GAs are extremely fast. Further, they are not very sensitive to network size [1,116]. The quality of solution (i.e., computed path) returned by NNs is constrained by their inherent characteristics. GAs are flexible in this regard. The quality of solution can be adjusted as a function of population. In addition, NN hardware is limited in size: it cannot accommodate networks of arbitrary size because of its physical limitation. GA hardware, on the other hand, scales well to networks that may not even fit within the memory. It is realized by employing parallel GA over several nodes. Therefore, GAs (especially hardware implementations) are clearly quite promising in this regard.

3.2 Existing GA-Based Approaches

Investigators have applied GAs to unicasting SP routing problem [57,67,79], multicasting routing problem [118,120], ATM bandwidth allocation problem [84], capacity and flow assignment problem [78], and the dynamic routing problem [108]. It is noted that all these problems can be formulated as some sort of a combinatorial optimization problem.

Munetomo's algorithm [79] is practically feasible in a wired or wireless environment. It employs variable-length chromosomes for encoding the problem. Crossing sites (i.e, crossover points) are the loci (viz., positions of nodes in a route) where identical genes (i.e., nodes) in both the chosen chromosomes (i.e., routes) are found at the same location. Thus, it leads to a situation in which only a few crossover sites are usable for exploring feasible solutions. In other words, crossover is totally dependent on positions: indeed, identical genes should occupy the same locus for crossover. The candidate crossing sites are called "potential crossing sites." A locus is selected randomly to act as an actual crossing site and to partially exchange chromosomes with the parent. In the mutation phase, a gene (i.e., the mutation node) is selected randomly from the chromosome. Another gene is selected randomly from the chromosomes connected directly to the mutation node, and a mutated chromosome (viz., alternative route) is generated by combining each partial chromosome (i.e., partial route) obtained by Dijkstra's algorithm. It must be noted that one partial route refers to a shortest path from the source node to the selected node and the other to a shortest path from the selected node to the destination node. But, the algorithm requires a relatively large population for an optimal solution due to the constraints on the crossover mechanism. Furthermore, it is not suitable for large networks or real-time communications since Dijkstra's algorithm has a prohibitive computational cost.

Inagaki et al. [57] proposed an algorithm that employs fixed (i.e., deterministic) length chromosomes. The chromosomes in the algorithm are sequences of integers and each gene represents a node ID, that is selected randomly from the set of nodes connected with the node corresponding to its locus number. All the chromosomes have the same (fixed) length. In the crossover phase, one of the genes from two parent chromosomes is selected at the locus of the starting node ID and put in the same locus of an offspring. One of the genes is then selected randomly at the locus of the previously chosen gene's number. This process is continued until the destination node is reached. The details of mutation are not explained in the algorithm. The algorithm requires a large population to attain an optimal or high quality of solution due to its inconsistent crossover mechanism. Some offspring may generate new chromosomes that resemble the initial chromosomes in fitness, thereby retarding the process of evolution.

There are several GAs that address different kinds of routing problems, such as multiple destination or multicasting routing problems [67, 118, 120]. Those approaches are beyond the scope of this investigation. However, the unicasting or one-destination algorithms such as the one proposed here can be extended in a straightforward manner to include them.

3.3 Proposed GA-based Routing Algorithm

The underlying topology of multihop networks can be specified by the directed graph $\mathbf{G} = (\mathbf{V}, \mathbf{A})$, where $\mathbf{V}$ is a set of $|\mathbf{V}|$ nodes (or vertices), and $\mathbf{A}$ is a set of its links (or arcs, edges) [8, 110]. There is a cost C_{ij} associated with each link (i, j). The costs are specified by the cost matrix $\mathbf{C} = [C_{ij}]$, where C_{ij} denotes a cost of transmitting a packet on link (i, j). Source and destination nodes are denoted by S and D, respectively. Each link has the link connection indicator denoted by I_{ij}, which plays the role of a chromosome map (i.e., masking) providing information on whether the link from node i to node j is included in a routing path or not. It can be defined as follows:

$$
I_{ij} = \begin{cases} 1, & \text{if the link } (i, j) \text{ exists in the routing path} \\ 0, & \text{otherwise.} \end{cases} \tag{3.1}
$$

It is obvious that all the diagonal elements of $\mathbf{I}$ must be zero. Using the above definitions, the SP routing problem can be formulated as a combinatorial optimization problem minimizing the objective function (Eq. (3.2a)) as follows:

minimize

$$
\sum_{\substack{i=S}}^{D} \sum_{\substack{j=S \\ j \neq i}}^{D} C_{ij} \cdot I_{ij} \tag{3.2a}
$$

subject to

$$
\sum_{\substack{j=S \\ j \neq i}}^{D} I_{ij} - \sum_{\substack{j=S \\ j \neq i}}^{D} I_{ji} = \begin{cases} 1, & \text{if } i = S \\ -1, & \text{if } i = D \\ 0, & \text{otherwise} \end{cases}
$$

and

$$
\sum_{\substack{j=S \\ j \neq i}}^{D} I_{ij} \begin{cases} \leq 1, & \text{if } i \neq D \\ = 0, & \text{if } i = D \end{cases}
$$

$$
I_{ij} \in \{0, 1\}, \text{ for all } i. \tag{3.2b}
$$

The constraint (Eq. (3.2b)) ensures that the computed result is indeed a path, without loops, between a source and a designated destination.

3.3.1 Chromosome Representation

A chromosome of the proposed GA consists of sequences of positive integers, which represent the IDs of nodes through which a routing path passes. Each locus of the chromosome represents an order of a node (indicated by the gene of the locus) in a routing path. The gene of the first locus is always reserved for the source node. The length of the chromosome is variable, but

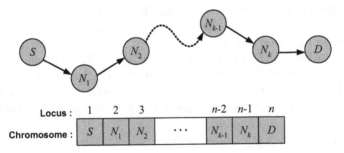

Fig. 3.1. Example of routing path and its encoding scheme.

it should not exceed the maximum length $|\mathbf{V}|$, where $|\mathbf{V}|$ is the total number of nodes in the network, since it never needs more than the total number of nodes to form a routing path. A chromosome (i.e., routing path) encodes the problem by listing up node IDs from its source node to its destination node based on topological information database (i.e., routing table) of the network. The information can be easily obtained and managed in real-time by routing protocols such as OSFP [72], DSR [80], and VCRP [2] in wired or wireless environments, but the detailed mechanisms or other controversial issues are beyond the scope of this study. It is noted that the topological information database of the network can be constructed easily and rapidly by such routing protocols.

An example of chromosome encoding from node S to node D is shown in Fig. 3.1. The chromosome, viz., routing path, is essentially a list of nodes along the constructed path, $(S \rightarrow N_1 \rightarrow N_2 \rightarrow \cdots \rightarrow N_{k-1} \rightarrow N_k \rightarrow D)$. In Fig. 3.1, n represents the total number of nodes forming a path.

The gene of the first locus encodes the source node, and the gene of second locus is randomly or heuristically selected from the nodes connected with the source node (S) that is represented by the front gene's allele. The chosen node is removed from the topological information database to prevent the node from being selected twice, thereby avoiding loops in the path. This process continues until the destination node is reached. Note that an encoding is possible only if each step of a path passes through a physical link in the network.

3.3.2 Population Initialization

Heuristic initialization may be beneficial to the SP routing problem because the topological information for computing the SP is already collected before the algorithm starts. However, the heuristic initialization may increase the complexity of the algorithm and lead to premature convergence (as described in Sect. 2.2). Consequently, random initialization is effected so that initial population is generated with the encoding method already explained in Sect. 3.3.1. Physically, the random initialization chooses genes (viz., nodes) from the topological information database in a random manner during the encoding process. It is possible that the algorithm encounters a node for which all of whose neighboring nodes have already been visited. In this case, the defective chromosome is refreshed and reinitialized. This may induce a subtle bias in which some partial paths are more likely to be generated. However, the meager bias does not significantly affect the performance of the algorithm. It is doubly so because the bias vanishes after evolving just a few generations.

3.3.3 Fitness Function

Fitness function must be defined with utmost care so that the quality of candidate solutions is accurately measured. Fortunately, the fitness function in the SP routing problem is obvious because the SP computation amounts to finding the minimal cost path. Therefore, the fitness function that involves computational efficiency and accuracy (of the fitness measurement) is defined as follows:

$$F_i = \left[\sum_{j=1}^{n_i-1} C_{g_i(j),g_i(j+1)} \right]^{-1} \tag{3.3}$$

where F_i represents the fitness value of the ith chromosome, n_i is the length of the ith chromosome, $g_i(j)$ represents the gene (i.e., node) of the jth locus in the ith chromosome, and C_{ij} is the link cost from node i to node j.

The fitness function of GAs is generally the objective function that requires to be optimized [38, 56, 67]. In a sense, the fitness function (Eq. (3.3)) can be thought of as fully reflecting the objective function (Eq. (3.2a)). The fitness function has a higher value when the fitness characteristic of the chromosome is better than others. In addition, the fitness function introduces a criterion for selection of chromosomes.

3.3.4 Genetic Operators

Great care must be exercised in designing genetic operators that lead GAs to the (globally) optimal solution, quickly, accurately, and reliably.

Selection

Among the ordinal selection schemes, tournament selection without replacement is promising as it is perceived to be effective in keeping the selection noise as low as possible (described in Sect. 2.2). Recall that the selection pressure of tournament selection increases with the tournament size s. In general, high selection pressure leads to premature convergence. Thus, the pairwise (i.e., $s = 2$) tournament selection without replacement is employed for the proposed GA: two chromosomes are picked and the one that is fitter is selected. However, the same chromosome should not be picked twice as a parent.

Crossover

Crossover examines the current solutions in order to find better ones. Physically, crossover in the SP routing problem plays the role of exchanging each partial route of two chosen chromosomes in such a manner that the offspring produced by the crossover represents only one route. This dictates selection of one-point crossover as a good candidate scheme for the proposed GA. One partial route connects the source node to an intermediate node, and the other partial route connects the intermediate node to the destination node. The crossover between two dominant parents chosen by the selection gives higher probability of producing offspring having dominant traits.

But the mechanism of the crossover is not the same as that of the conventional one-point crossover. In the proposed scheme, two chromosomes chosen for crossover should have at least one common gene (i.e., node) except for source and destination nodes, but there is no requirement that they be located at the same locus. That is, the crossover does not depend on the position of nodes in routing paths. Figures 3.2(a) and (b) show the pseudo-code and an example of the crossover procedure, respectively.

As shown in Fig. 3.2(b), a set of pairs of nodes which are commonly included in the two chosen chromosomes but without positional consistency is formed first, viz., (3,2) and (5,4). Such pairs are also called "potential crossing sites." Then, one pair (3,2) is randomly chosen and the locus of each node becomes a crossing site of each chromosome. The crossing points of two chromosomes may be different from each other. This is in contrast to the scheme adopted in Munetomo's algorithm [79]. Each partial route is exchanged and assembled, eventually leading to two new routes. It is possible that loops are formed during crossover. In this regard, a simple countermeasure must be prepared with a view to avoiding degenerating the rate of convergence and the quality of solution. Of course, such chromosomes (viz. routes with loops) will gradually be weeded out in the course of a few generations because the traits of those chromosomes drive fitness values from bad to worse. Repair and penalty functions are the usual countermeasures. It is described in Sect. 3.3.5.

Mutation

The population undergoes mutation by an actual change or flipping of one of the genes of the candidate chromosomes, thereby keeping away from local optima. Physically, it generates an alternative partial route from the mutation node to the destination node in the proposed GA. A topological information database is utilized for the purpose. Of course, mutation may induce a subtle bias for reasons indicated earlier (Sect. 3.3.2). However, this bias can be ignored. This is explained below.

First, mutation leads to an infinitesimal increase in the probability of inducing the bias. Second, selection and crossover strongly influence the way this

PARAMETERS.
C_1, C_2 : Input chromosomes, $\widetilde{C}_1, \widetilde{C}_2$: Output chromosomes
n_1, n_2 : Length of chromosomes C_1, C_2

```
/*Find the potential crossing sites*/
for i:=1 to n₁ do
    for j:=1 to n₂ do
        /*If a node is common to both chromosomes*/
        if C₁[i]==C₂[j] then
            /*Construct a set of potential sites*/
            sₚ[k]:=(i,j);
            k++;

/*Randomly choose a crossing site*/
sc:=choose_random(sₚ);
C̃₁:=C₁[1 : sc(1)]//C₂[sc(2) + 1 : n₂];  /*1st exchange*/
C̃₂:=C₂[1 : sc(2)]//C₁[sc(1) + 1 : n₁];  /*2nd exchange*/
```

(a) Pseudo-code of the crossover.

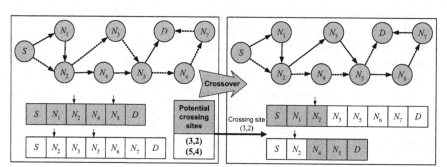

(b) Example of the crossover procedure.

Fig. 3.2. Overall procedure of the crossover.

bias operates. Indeed, its (bias') harmful effects vanish almost completely. Furthermore, the bias, whenever it helps in searching an optimal solution, may not induce any harmful effect at all.

Figure 3.3 shows the overall procedure of the mutation operation. As can be seen from Fig. 3.3(b), in order to perform a mutation, a gene (i.e., node N_2) is randomly selected first from the chosen chromosome ("mutation point"). One of the nodes, connected directly to the mutation point, is chosen randomly as the first node of the alternative partial route. The remaining procedure has been given in Sects. 3.3.1 and 3.3.2.

However, nodes already included in an upper partial route should be deleted from the database so as not to include the same node twice in the new routing path. The upper partial route represents the surviving portion

PARAMETERS.
C: Input chromosome, $\widetilde{C}$: Output chromosome
T: Topological information database

```
/*Randomly choose a node as a mutation point*/
sₘ:=choose_random(C);
/*Delete the nodes of upper partial route from T*/
delete(T,C,sₘ);
/*Put the upper partial route to the mutation chromosome*/
C̃:=C[1 : sₘ];

/*Construct the remaining mutation chromosome*/
while(1)
    /*Randomly choose a node and then delete the node from T*/
    C̃[sₘ + 1]:=choose_random_delete(T,C̃[sₘ]);
    if C̃[sₘ + 1]==D then
        break;
    sₘ++;
```

(a) Pseudo-code of the mutation.

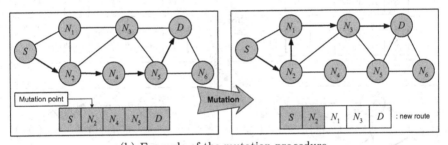

(b) Example of the mutation procedure.

Fig. 3.3. Overall procedure of the mutation.

of the previous route after mutation; it is the partial chromosome stretching from the first gene to the intermediate gene at the mutation point.

3.3.5 Repair Function

As mentioned earlier, crossover may generate infeasible chromosomes that violate the constraints (Eq. (3.2b)), generating loops in the routing paths. It must be noted that none of the chromosomes of the initial population or after the mutation is infeasible because when once a node is chosen, it is excluded from the candidate nodes forming the rest of the path.

Due to critical drawbacks of the penalty method – the difficulty of devising an appropriate penalty function and the reproduction of infeasible chromosomes at the price of some feasible ones, the repair method is employed in the

PARAMETERS.
C: Input chromosome, $\widetilde{C}$: Output chromosome
n: Length of chromosome C

```
/*Find and eliminate a loop*/
for i:=1 to n
    for j:=1 to n
        if C[i]==C[n−j] then
            C̃:=C[1 : i]//C[n − j + 1 : n];
```

(a) Pseudo-code of the repair function.

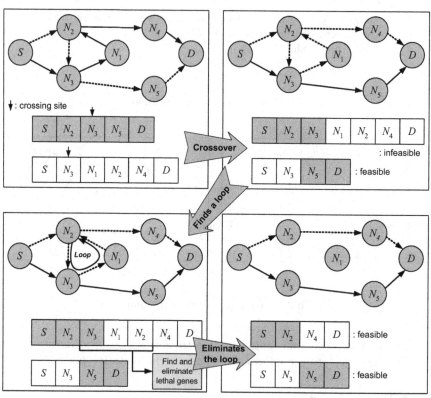

(b) Example of the repair function.

Fig. 3.4. Overall procedure of the repair function.

proposed GA. Fortunately, the mechanism that eliminates the lethal genes that form loops can cure all the infeasible chromosomes. The repair function finds and eliminates loops in a routing path without unduly increasing computational costs.

The proposed repair function is described in Fig. 3.4(a) and an example is shown in Fig. 3.4(b). In Fig. 3.4(b), one of the offspring produced after crossover becomes infeasible because the new route contains the loop $N_2 \rightarrow N_3 \rightarrow N_1 \rightarrow N_2$. The repair function detects the loop by a simple search described in Fig. 3.4(a). The function is linear in n, the chromosome length. After that, the lethal genes (forming a loop) that violate the constraint condition are deleted; those nodes are $\{N_3, N_1, N_2\}$[1] in this example.

3.3.6 Population Size

In the GA run, a proper setting of the population size is the most crucial issue. To find better solutions, the population size must be increased as much as possible; however, it may result in unacceptably slow convergence. In Sect. 2.3, a practical population-sizing model has been developed. It explicitly tells us of the relationships among the size of the population, the quality of solution, the cardinality of the alphabet, and other factors of GAs. One can compute an adequate population size by making use of Eq. (2.14). A detailed investigation of the validity and usefulness of the population-sizing model to the SP routing problem is treated in Sect. 3.4.3.

3.4 Experiments and Discussion

In this section, the proposed GA is compared with Munetomo's [79] and Inagaki's [57] algorithms through computer simulations. As described in Sect. 3.3.4, the proposed GA employs pair-wise tournament selection (i.e., tournament size $s = 2$) without replacement. In all the experiments, the crossover and mutation probabilities are set to 1.0 and 0.05, respectively.[2] Each experiment is terminated when all the chromosomes have converged to the same solution. A convergence test for termination is performed before applying mutation as otherwise uninvited evolution may follow due to the placement of the mutation operator. However, this strategy does not affect the results.

Each solution is compared with Dijkstra's SP [110] solution. In other words, Dijkstra's algorithm provides a reference point. Furthermore, the accuracy and the scalability of the population-sizing model are also verified through simulation studies.

3.4.1 Results for a Fixed Network with 20 Nodes

The simulation studies involve the deterministic, weighted network topology with 20 nodes depicted in Fig. 3.5(d). The bold line shows an optimal path.

---

[1] One of two N_2 must be taken regardless of its order.
[2] In general, they come under a set of typical values.

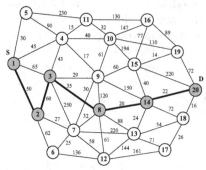

(a) Result of the Munetomo's algorithm (total path costs: 187).

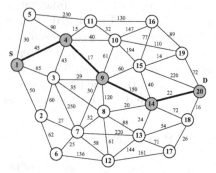

(b) Result of the Inagaki's algorithm (total path costs: 234).

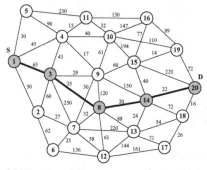

(c) Result of the Proposed algorithm (total path costs: 142).

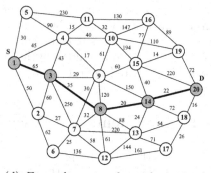

(d) Example network with optimal path in bold line (optimal costs: 142).

Fig. 3.5. Comparison results for the paths found by algorithms.

With a view to focusing exclusively on fair comparison of algorithms on the basis of performance, the population size is taken to be the same as the number of nodes in the network. Population size and its influence on quality of solution are investigated later (Sect. 3.4.3).

Fig. 3.5 shows the shortest paths (viz., bold lines) found by the algorithms for the indicated source-destination pair. It is seen that the path computed by the proposed algorithm coincides with that found by Dijkstra's algorithm. The latter is known to always return the optimal shortest path. Munetomo's and Inagaki's algorithms, on the other hand, settle for a suboptimal path.

Figure 3.6 compares objective function values returned by the algorithms. In the figure, the objective function value represents the sum of link costs (i.e., total path cost) normalized by the maximum link cost in the network. It is seen that the proposed GA exhibits the fastest rate of convergence because the number of generations up to convergence is the smallest. The algorithm

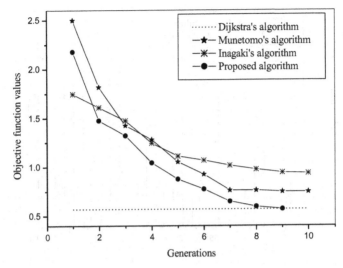

Fig. 3.6. Convergence property of each algorithm.

converging through smaller generations has better convergence performance because all the algorithms have the same population size in the experiment. In general, however, convergence performance must be compared with the average number of fitness function evaluations until the GAs reach equal quality of solutions [22]. A detailed explanation will be given in Sect. 3.4.2. In addition, it indeed converges to a value that is exactly the same as the Dijkstra-value (i.e., the optimal route), notwithstanding the somewhat inherent initial disadvantage.

3.4.2 Results for Random Networks

In this section, we verify that the previous results concerning the quality of solution and the convergence speed hold for all kinds of problems (i.e., networks types and scales). Networks with 15–50 nodes, and randomly assigned (normalized) link costs were investigated. As described earlier (see Sect. 3.1), mobile *ad hoc* networks provide acceptable targets. Applications include military battlefield (e.g., moving platoon or company), rescue missions, conference room and so on [2,80]. They involve networks with sizes that range from small to medium (e.g., tens of nodes). Simulations reflect this practical reality. A possible implication is that the proposed algorithm scales well to larger networks.

First, the quality of solution (i.e., route optimality) for each GA is investigated. The route optimality is defined as a percentage of the number of times the GA finds the global optimum (i.e., the shortest path). The route failure ratio is the inverse of route optimality. It is asymptotically the probability that the computed route is not optimal, because it is the relative frequency

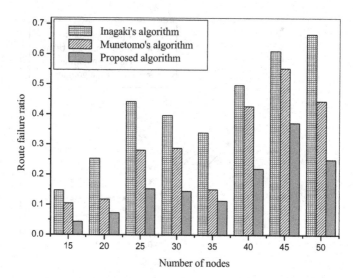

Fig. 3.7. Comparison results of the quality of solution for each algorithm.

Table 3.1. Performance comparison on the quality of solution.

Performance measure		Inagaki's algorithm	Munetomo's algorithm	Proposed algorithm
Route failure ratio (RFR)	μ_{RFR}	0.4195	0.2959	0.1712
	σ_{RFR}	0.1745	0.1670	0.1067

of route failure. The population size of each GA is also taken to be the same as the number of nodes in the networks. A total of 1000 random network topologies were considered in each case.

The quality of solutions of the algorithms is compared in Fig. 3.7. From the figure, we can see that the quality of the solution of the proposed GA is much higher than that of the other algorithms. In case of 30 nodes, for example, the proposed GA outperforms Inagaki's GA and Munetomo's GA with *prob.* < 0.26 and *prob.* < 0.15, respectively. The results are collected in Table 3.1. The proposed GA attains a 0.1712 route failure ratio (viz., 82.88% route optimality) with a population size equal to the number of nodes in the networks. The proposed GA is better than Inagaki's GA and Munetomo's GA with *prob.* < 0.25 and *prob.* < 0.13, respectively. Meanwhile, standard deviation of route failure ratio (i.e, probability) for the proposed GA amounts to 0.1067 comparing favorably with 0.1745 for Inagaki's GA and 0.167 for Munetomo's GA. It means that the proposed algorithm retains its robustness amidst changing network topologies with regard to the quality of solution (i.e.,

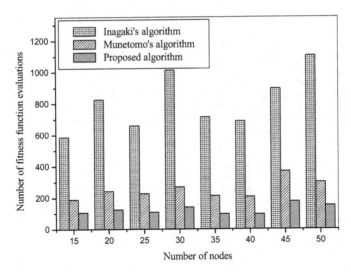

Fig. 3.8. Comparison results of the rate of convergence for each algorithm.

route optimality). The benefits accrue from the compound effects of effective search capability of the crossover and the diversity maintenance by mutation.

Second, the convergence speed of every GA is investigated. Convergence performance is investigated in terms of the average number of fitness function evaluations. Cantú-Paz [22] suggests the total execution time (i.e., the number of fitness function evaluations) required to find a solution of the same average quality as a fair comparison criterion. In other words, the number of fitness function evaluations can directly measure the dominance of convergence performance only if all the GAs converge to solutions with identical quality. Unfortunately, finding the exact population size for a particular quality of solution for each algorithm is very difficult. However, determining the population size with certain constraints is relatively easy. We can determine the population size for each algorithm by exhaustive search for each algorithm so as to achieve almost the same quality of solution (viz., route optimality). The proposed algorithm seems to have the most satisfactory performance and Inagaki's the least. The reason is not far to seek: the proposed algorithm involves the smallest number of fitness function evaluations. That means faster convergence.

Networks with 15–50 nodes, and randomly assigned link costs were also studied. The results in respect of number of fitness function evaluations are shown in Fig. 3.8. From the figure, we can see that the convergence rate of the proposed algorithm is much higher than that of the other algorithms in every case, because the average number of fitness function evaluations needed to reach similar quality of solution (i.e., maximum difference is about 4%) is smaller than any other algorithm. In case of 30 nodes, for instance, the proposed GA is faster than Inagaki's GA and Munetomo's GA with *prob.* <

Table 3.2. Performance comparison on the rate of convergence.

Performance measure		Inagaki's algorithm	Munetomo's algorithm	Proposed algorithm
Achieved route failure ratio		0.2438	0.229	0.2014
Number of fitness function evaluations	μ_{eval}	810.5440	251.9105	122.3158
	σ_{eval}	181.9882	58.9393	31.6627

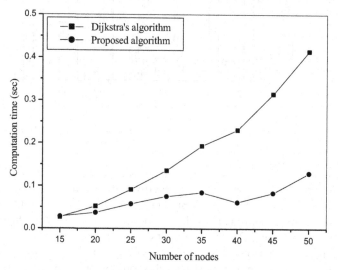

Fig. 3.9. Computation time between Dijkstra's and the proposed algorithms.

0.87 and *prob.* < 0.48 (i.e., 7.27 times and 1.92 times), respectively. The results are collected in Table 3.2.

From Table 3.2, it can be seen that the proposed GA converges to a solution with about 0.2 route failure ratio (viz., 80% route optimality) in about 122 fitness function evaluations. The convergence speed of the proposed GA is superior to that of Inagaki's GA and Munetomo's GA with *prob.* < 0.85 and *prob.* < 0.52 (i.e., 6.63 times and 2.06 times), respectively. It is noted that such improvement serves as a lower bound of convergence gain because the proposed GA still attains better quality of solution than other algorithms. Furthermore, standard deviation of fitness function evaluations for the proposed GA is about 32, while it is about 182 for Inagaki's GA and about 59 for Munetomo's GA. It also implies that the proposed GA is indeed insensitive to network topologies, as far as convergence is concerned.

In order to further compare the convergence performance of the proposed GA with that of Dijkstra's algorithm, direct (real) computation time obtained from previous experiments given in Fig. 3.7 is presented in Fig. 3.9. The average computation time of Dijkstra amounts to 0.14 second and that of the proposed GA is 0.067 second. The proposed GA is faster than Dijkstra's algorithm with *prob.* < 0.522, although it may confront route failure situation with *prob.* < 0.1712. The computation time of the proposed GA does not increase significantly with the network size while it does in case of Dijkstra's algorithm. It is noted that both the algorithms are, *per se*, inadequate for real-time communications in mobile *ad hoc* networks. However, the proposed GA, with its extremely fast hardware counterpart, passes the test (see Sect. 3.1).

3.4.3 Experimental Verification of the Population-Sizing Model

In this section, we verify that the generalized version of the population-sizing model of Eq. (2.14) fairly accurately predicts the quality of solutions computed by the proposed GA.

Deciding the Parameters in the SP Routing Problem

In Sect. 2.3, the practical population-sizing model was given by

$$N = -\frac{\chi^k}{2}\ln(\alpha)\left(\frac{\chi^k - 1}{2}\sqrt{\pi m'} + 1\right). \tag{3.4}$$

To apply it to the SP routing problem, the parameters have to be determined. As χ is the average cardinality of the alphabet, it can be found by the average link connectivity in the network. In other words, the parameter χ physically represents the average number of nodes that each gene can take. As mentioned before, the parameter α is the GA failure (i.e., route failure) probability[3]. The average order k of BBs can be modeled as a linear combination of one-max function and deceptive function in the SP routing problem. That is,

$$k = \sum_{x=1}^{|\mathbf{V}|}(c_x \cdot x) \tag{3.5}$$

where $|\mathbf{V}|$ is the number of nodes in the networks, and c_x, the weighted average coefficient is a domain-dependent parameter. The sum of all the coefficients is 1.

In Eq. (3.5), we get the one-max problem when $x = 1$, and the deceptive problem when $x > 1$. In addition, the total number of collateral noise sources m' is $(m-1)$, where m is the average number of BBs. In the routing problem, m is calculated as n/k, where n is the average length of chromosomes that is defined by the number of nodes whose average cost is not greater than that of the overall network. Of course, the BBs may be inherently interdependent. However, the average number of BBs (n/k) will be a reasonable approximation

---

[3] It denotes the probability that the computed route is not optimal.

from a statistical point of view, if an accurate average order k is used. This is explained below.

The possibility of finding a shortest path will be quite high when each node chooses a lowest/best-cost node among its own neighbors as a forward node in a route (as happens in a greedy algorithm). If the shortest path is always found in this manner, the problem can be modeled as the one-max problem. But, a globally optimal path is possible even when locally non-optimal selections are made. It is well known that locally optimal selections may be misleading. These features reflect on the interdependence among BBs. The idea is to make an attempt to spread this potential (to mislead) over the average length of chromosomes, and thereby weaken it. The chromosomes can then be modeled as independent BBs.

The average order k is a unique domain-dependent variable not yet specified. Assuming that it is likely that an average order of more than two is very rare, the parameter k can be approximated by a two-term weighted average as follows:

$$k \approx \sum_{x=1}^{2} c_x \cdot x. \qquad (3.6)$$

The reason for this assumption is explained below. When k is 2, half the nodes in a route choose lower/worse nodes among their own neighbors as forward nodes in order to build an optimal path (i.e., quite misleading). Thus, $k > 2$ can be ignored since this situation itself (i.e., $k = 2$) is relatively rare in practice. Determining the coefficients is a very difficult problem. They are also sensitive to network size and topology. We observed that plotting the average deceptive size against the number of nodes on a log-log scale results in an almost straight line, which means that the coefficient of the average order 2, c_2, can be approximated with a general power-law equation [22] as follows:

$$c_2 = A \cdot |\mathbf{V}|^B. \qquad (3.7)$$

Here, A and B are domain-dependent constants. The value of A and B can be computed by transcendental cognition as follows:

$$A = 10^{-2} \cdot (1 - \alpha)^2 \quad \text{and} \quad B = 1.0. \qquad (3.8)$$

Therefore, the average order may be calculated as follows:

$$k = 1 \cdot c_1 + 2 \cdot c_2 = 1 + c_2 = 1 + 10^{-2} \cdot (1 - \alpha)^2 \cdot |\mathbf{V}|. \qquad (3.9)$$

From Eq. (3.9), we can see that the average order k is around 1 if the network does not have a large number of nodes. In that case, the probability of disruption of the BBs by crossover is very small. It is noted that if the average order k becomes large, the probability becomes large too and the population size might be affected. As mentioned earlier (see Section 2.3.4), however, the increasing average order does not strongly induce any increment in the population size if one- or two-point crossover is exploited [45].

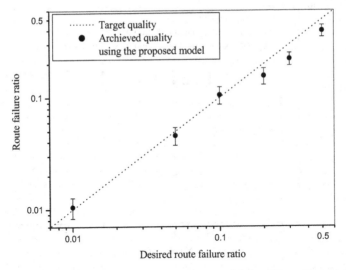

Fig. 3.10. Validity of the population-sizing model for SP routing problem.

Experimental Results

For an accurate assessment, all the results must be averaged over 10000 random network topologies, because the best quality of solution in the experiment guarantees a 10^{-2} route failure ratio (viz., 99% route optimality). In other words, we must experiment until the number of suboptimal solutions is at least 100; thus, 10000 network topologies are needed for supporting up to 10^{-2} route failure ratio.[4] In this respect, the experiments are run ten-times with a total of 1000 random network topologies with 15–50 nodes, and randomly assigned link costs (normalized).

Experiments concerning the population-sizing model are summarized in Fig. 3.10. In that figure, the dotted line is the target route failure probability and the symbol is the experimental result with the population size obtained by Eq. (3.4). From the figure, we can see that the model accurately predicts a population size that is adequate for reaching the desired (target) solution on the order of route failure probability of 0.1 (i.e., better than 90% optimality), and satisfactorily estimates the population size as an upper bound in cases of worse route failure probabilities. Thus, the model can be used for determining the population size for a desired quality of solution.

On the other hand, the scalability of the population-sizing model with regard to the number of nodes (i.e., network size) is investigated only in two cases, viz., 0.01 (99%) and 0.1 (90%) route failure probabilities (route optimality). The results of the experiments are shown in Fig. 3.11. It can be seen

---

[4] The reason for employing this methodology lies in ensuring the obtained results by observing sufficient samples.

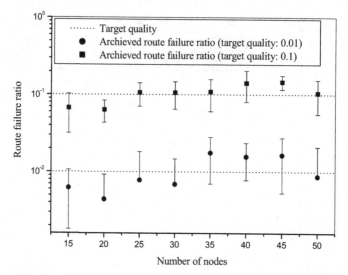

Fig. 3.11. Scalability of the population-sizing model for SP routing problem.

that the quality of solution is not strongly influenced by the number of nodes since the variation of the quality of solutions is about 5% (with regard to route optimality). It means that the model is scalable at 0.01 and 0.1 route failure ratio.

As a result, it may be inferred that the practical population-sizing model is scalable, a characteristic that is quite important in practical problems.

3.5 Summary

This chapter presented a GA for solving the SP routing problem. The crossover and the mutation operations work on variable-length chromosomes. The crossover is simple and independent of the location of the crossing site. Consequently, the algorithm can search the solution space in a very effective manner. The mutation introduces, in part, a new alternative route. In essence, it maintains the diversity of population thereby avoiding local traps. Infeasible solutions (i.e., chromosomes) have also been dealt with without unduly compromising on computational requirements.

Simulation studies show that the algorithm is indeed insensitive to variations in network topologies in respect of both route optimality (i.e., quality of solution) and convergence speed. Experimental results show that the quality of solution is better than those of other algorithms. Indeed, 0.1712 route failure ratio (viz., 82.88% route optimality) is attained with the population size that is equal to the number of nodes in the networks. The route failure (i.e., optimality) performance of the proposed GA was better than those of Inagaki's GA and Munetomo's GA with *prob.* < 0.25 and *prob.* < 0.13, respectively.

Furthermore, convergence was seen to occur after about 122 fitness function evaluations with up to about 0.2 quality of solution (viz., 80% route optimality). The convergence performance of the proposed GA was better than those of Inagaki's GA and Munetomo's GA with *prob.* < 0.85 and *prob.* < 0.52, respectively. In addition, the actual computation time of the proposed GA was shorter than that of Dijkstra's algorithm. However, hardware implementation may be necessary for applications involving real-time services in a dynamic network topology.

The accuracy of the practical population-sizing model was also verified, in context of the SP routing problem, with experiments by specifying different solution qualities ranging from 0.01 to 0.5 route failure probability (that is, 50~99% route optimality). The results showed that the predictions of the model are acceptable in the practically operational region which is better than 0.1 route failure probability (i.e., 90% route optimality). It can be used as an upper bound elsewhere. Furthermore, the model scales with problem difficulty (i.e., network size).

The proposed algorithm can search the solution space effectively and speedily compared with other extant algorithms. The population-sizing model appears to be a conservative tool to determine a population size in the routing problem.

4

Elitist Compact Genetic Algorithms

This chapter describes two elitism-based compact genetic algorithms (cGAs) – *persistent elitist compact genetic algorithm* (pe-cGA), and *nonpersistent elitist compact genetic algorithm* (ne-cGA). The aim is to design efficient compact-type GAs by treating them as simple *estimation of distribution algorithms* (EDAs) for solving difficult optimization problems without compromising on memory and computation costs. Difficult problems have the following characteristics: 1) full deception, 2) interdependence (of decision variables), 3) multimodality, and 4) symmetry. The idea is to deal with issues connected with lack of memory – inherent disadvantage of cGAs – by allowing a selection pressure that is high enough to offset the disruptive effect of uniform crossover. The point is to properly reconcile the cGA with elitism. The pe-cGA finds a near optimal solution (i.e., a winner) that is maintained as long as other solutions (i.e., competitors) generated from probability vectors (PVs) are no better. It attempts to adaptively alter the selection pressure in accordance with the degree of problem difficulty by employing only the pairwise tournament selection strategy. Moreover, it incorporates a model that is equivalent to the (1+1) evolution strategy (ES) with self-adaptive mutation. The pe-cGA, apart from providing high performance, also reveals the hidden connection between EDAs (e.g., cGA) and ESs (e.g., (1+1)-ES). The ne-cGA further improves the performance of the pe-cGA by avoiding strong elitism that may lead to premature convergence (by incorporating elitism in a restricted manner). The ne-cGA offers all the benefits of the pe-cGA. Restricted elitism also plays a role in arresting the rapid degeneration of genetic diversity (i.e, diversity maintenance). This chapter also proposes an analytic model for investigating convergence enhancement, viz., *speedup*.

The chapter is organized as follows. Section 4.1 briefly describes a family of cGAs. Section 4.2 explains the original cGA and provides an overview of elitism. In Section 4.3, the proposed elitism-based compact GAs for efficiently solving difficult problems are described. The (analytic) speedup model is presented in Sect. 4.4. The results of experiments on several test functions/problems and Ising Spin-Glasses (ISG) systems (a real-world applica-

Chang Wook Ahn: *Advances in Evolutionary Algorithms: Theory, Design and Practice*, Studies in Computational Intelligence (SCI) **18**, 45–83 (2006)
www.springerlink.com   © Springer-Verlag Berlin Heidelberg 2006

tion) can be found in Sect. 4.5. The chapter concludes with a summary of the results in Sect. 4.6.

4.1 A Family of Compact Genetic Algorithms

Of all the issues connected with genetic algorithms (GAs) – such as population size, genetic operators (e.g., selection, crossover, and mutation), and encoding methods, etc., – the population size that guarantees an optimal solution quickly enough has been a topic of intense research [3,39,40,45,49,101]. This is because large populations generally result in better solutions, but at increased computational costs and memory requirements. Goldberg and Rudnick [39] developed the first population-sizing model based on the variance of fitness. They further required the equation to permit accurate statistical decision making in the presence of competing building blocks (BBs) [40]. However, if wrong BBs are chosen in the first generation, the GAs will never recover [45,101]. Extending the decision model in [40], Harik *et al.* [45] exploited the similarity between the gambler's ruin problem and the selection mechanism of GAs for determining an adequate population size that guarantees a solution with the desired quality. Furthermore, the analytic model started from the assumption that the fitness values of a pair of chromosomes can be ordered. This effectively implies tournament selection without replacement. Moreover, Ahn and Ramakrishna [3] further enhanced and generalized the population sizing model in [45] so as to dispense with any problem dependent information such as signal or collateral noise of competing BBs. While attempting to understand the real importance of population in evolutionary algorithms (EAs), He and Yao [49] showed that the introduction of population increases the first hitting probability, so that the mean first hitting time is shortened.

Based on the results in [45], Harik *et al.* [46] proposed the *compact GA* (cGA) as an (simple) estimation of distribution algorithm (EDA) that generates the offspring in line with the estimated probabilistic model of parent population instead of using traditional recombination and mutation operators [64,115]. The cGA represents the population as a probability (distribution) vector (PV) over the set of solutions and operationally mimics the order-one behavior of simple GA (sGA) with uniform crossover using a small amount of memory. Therefore, the cGA can be a welcome too in memory-constrained applications such as multicasting routing and resource allocation problems in the emerging field of wireless networks. When confronted with easy problems (e.g., continuous-unimodal problems) involving lower order BBs, the cGA can achieve solutions of comparable quality with approximately the same number of fitness evaluations as the sGA with uniform crossover [46]. However, the cGA does not provide acceptable solutions to difficult problems (e.g., deceptive problems or multimodal problems), because it does not have the memory to retain the required knowledge (e.g., decision error, linkage information of genes) about non-linearity of the problems [46]. These problems involve higher order BBs. It is noted that most practical applications may come within the purview of difficult optimization problems because they usually have many local optima (i.e., multimodal) and the genes are inter-dependent in general.[1] It follows that the cGA may not be effective in solving real-world problems. In order to obtain better solutions to such difficult problems, the cGA should exert a higher selection pressure. This, in turn, increases the survival probability of higher order BBs, thereby preventing loss of the best solution found thus far. In other words, higher selection pressure can play the role of memory. Therefore, it can take care of a finite number of decision errors and some linkage information of genes. Selection pressure of the cGA can be increased with easy by creating a larger tournament size [46]. However, this scheme requires additional but insignificant memory that is proportional to tournament size. Furthermore, it is difficult to precompute the (average) order of BBs in practical applications. Even if the order of BBs can be found (or is known) in advance, the tournament size that provides a selection pressure that is high enough to compensate for the highly disruptive effects of uniform crossover cannot be determined precisely. Although Harik *et al.* [46] investigated the relationship between tournament size and selection pressure in the cGA by employing a global schema theorem, the relation was verified only in the context of a specific problem involving concatenation of 10 copies of a 3-bit, fully deceptive function with deceptive-to-optimal ratio of 0.7. This underscores the point that the relationship may only be partially satisfied. This is because the tournament size is closely related to not only the order of BBs but also to other factors such as deceptive-to-optimal ratio and collateral noise. The problem with Harik's result is demonstrated through experiments in Sect. 4.5.2, but a detailed investigation is beyond the scope of this work.

Furthermore, Harik *et al.* [44] proposed an extended compact GA (ecGA) for solving difficult problems such as fully deceptive problems by combining a greedy marginal product model (MPM) search algorithm with a minimal description length (MDL) search model. Although the MPMs are similar to the PV of cGA with regard to the products of marginal distributions on a partition (i.e., a BB) of the genes, they can provide a direct linkage map with each partition separating tightly linked genes [100]. Thus, the ecGA can find better solutions to difficult problems with a smaller number of function evaluations (than the sGA). However, it requires more memory and carries higher computational costs per function evaluation.

Baraglia *et al.* [15] proposed a hybrid heuristic algorithm that combines cGA with an efficient Lin-Kernighan (LK) local search algorithm, the so called cGA-LK. The aim of cGA-LK is to deal with difficult order-k ($k > 1$) optimization problems such as the traveling salesman problem (TSP) without requiring larger memory than the existing cGA. The cGA-LK exploits the cGA in order to generate high quality solutions (to TSP), which are then refined with the LK local search algorithm. The refined solutions are in turn

---

[1] It is difficult to model the problems as the combination of lower order BBs.

exploited further with a view to improving the quality of the simulated population (i.e., the probabilities of PV). In this way, it achieves a performance that is better than is possible with sGA and cGA in terms of quality of solutions. However, the algorithm may incur an unacceptably high computational cost because it employs the complex LK local search algorithm.

In the same context, Hidalgo *et al.* [52] devised a hybrid algorithm for Multi-FPGA partitioning. The mechanism that combines the existing cGA and a random local search algorithm is quite similar to that of cGA-LK. Every time a certain number of epochs elapses, the best individual competes with a new individual found by local search. If the new individual has a higher fitness, then the PV (of cGA) is updated by its traits.

4.2 Compact Genetic Algorithm and Elitism

This section provides background information on cGA and elitism.

4.2.1 Compact Genetic Algorithm

In the community of genetic and evolutionary computation, *estimation of distribution algorithms* (EDAs), also known as *probabilistic model building genetic algorithms* (PMBGAs), have attracted due attention of late [64], [90].[2] Being one of the simplest EDA, the cGA manages its population as a PV over the set of solutions (i.e., only models its existence), thereby mimicking the order-one behavior of the sGA with uniform crossover using a small amount of memory [15, 46].

Figure 4.1 is the pseudo-code of the cGA. The elements of PV, viz., $p_i \in [0,1]$, $\forall i \in \{1, \cdots, n\}$, where n is the number of genes (i.e., the length of the chromosome), measures the proportion of "1" alleles in the ith locus of the simulated population [15, 46]. The PV is initially assigned 0.5 to represent a randomly generated population. In every generation (i.e., iteration), competing chromosomes are generated on the basis of the current PV, and their probabilities are updated to favor a better chromosome (i.e., winner). It is noted that the generation of chromosomes from PV simulates the effects of crossover that leads to a decorrelation of the population's genes. In a simulated population of size N, the probability p_i is increased (decreased) by $1/n$ when the ith locus of the winner has an allele of "1" ("0") and the ith locus of the loser has an allele of "0" ("1"). If both the winner and the loser have the same allele in each locus, then the probability remains the same. This scheme is equivalent to (steady-state) pairwise tournament selection [46]. The cGA is terminated when all the probabilities converge to 0.0 or 1.0. The convergent PV itself represents the final solution. It is seen that the cGA requires $n \cdot log_2(N+1)$ bits of memory while the sGA requires $n \cdot N$

---

[2] A detailed investigation about EDAs can be found in Sect. 5.1.

Compact Genetic Algorithm

PARAMETERS.
N: Population size, n: Chromosome length, p: Probability vector

STEP 1. Initialize probability vector
```
for j:=1 to n do
    p[i]:=0.5;
```
STEP 2. Generate two chromosomes from the probability vector
```
a:=generate(p);    b:=generate(p);
```
STEP 3. Let them compete
```
winner,loser:=compete(a,b);
```
STEP 4. Update the probability vector toward the better one
```
for i:=1 to n do
    if winner[i] ≠ loser[i] then
        if winner[i]==1 then p[i]:=p[i]+1/N;
        else p[i]:=p[i]-1/N;
```
STEP 5. Check if the probability vector has converged
```
for i:=1 to n do
    if p[i]>0 and p[i]<1 then
        go to STEP 2;
```
STEP 6. The probability vector represents the final solution

Fig. 4.1. Pseudo-code of compact genetic algorithm.

bits [46]. Thus, a large population size can be effectively exploited without unduly compromising on memory requirements [15].

On the other hand, the cGA simulates higher selection pressure to solve problems with higher order BBs. Selection pressure of the cGA can be increased by replacing STEPS 2–4 (in Fig. 4.1) with the procedures described below [46].

First, s chromosomes are generated from the PV. The best among them is found. Second, the best chromosome competes with the other $(s-1)$ chromosomes and the PV is updated on the way. However, it requires bothersome information such as the order of BBs and the tournament size[3]. Since such information may not be available in practice, the cGA may not be all that useful.

4.2.2 Elitism

As an operational characteristic of GAs, elitism provides a means for reducing genetic drift by ensuring that the best chromosome(s) is allowed to pass/copy their traits to the next generation [32, 94]. Genetic drift is used to

---

[3] It closely adjusts the selection pressure that is sufficient to combat the highly disruptive effects of uniform crossover.

explain/measure stochastic changes in gene frequency through random sampling of the finite population [95]. Some genes of chromosomes may turn out to be more important to the final solution than others [94]. When the chromosomes representing decision variables that have a reduced "salience" to the final solution do not experience sufficient selection pressure, genetic drift may be stalled. Therefore, it is important to maintain adequate selection pressure, as demanded by the application, in order to avoid this phenomenon [94]. In other words, the arrest of genetic drift reflects the failure to exert adequate selection pressure by increasing the tournament size or by some form of elitism.

Since elitism can increase the selection pressure by preventing the loss of low salient genes of chromosomes due to inadequate (i.e., deficient) selection pressure, it improves the performance with regard to optimality and convergence of GAs in many cases. However, the degree of elitism should be adjusted properly and carefully because high selection pressure may lead to premature convergence [32].

4.3 Elitism-Based Compact Genetic Algorithms

This section describes the *persistent elitist compact GA* (pe-cGA), and the *nonpersistent elitist compact GA* (ne-cGA). They combine the existing cGA with elitism in an effective manner. The major objective is to improve the quality of solution and the rate of convergence (to the global optimum) with acceptable memory and computational costs. Since the cGA operates on each gene independently, it may lose linkage information. As a consequence, the cGA may not be able to solve difficult problems, especially those involving higher order BBs (e.g. deceptive problems).

4.3.1 Persistent Elitist Compact Genetic Algorithm

In Sect. 4.2.1, we found that the selection pressure of cGA should be proportional to the degree of difficulty of problems for efficiently solving them. In other words, a more difficult problem requires a higher selection pressure for finding a better solution. This is because higher selection pressure offsets the disruptive effects of uniform crossover (i.e., it carries partial knowledge about the gene's correlation such as the linkage information), thereby encouraging convergence to a better solution. Although the selection pressure of the cGA can be increased by creating a larger tournament size, it requires additional but by no means significant memory and problem-dependent information that is not generally available in real-world problems. Even if such information is available, computation of the necessary tournament size that builds a selection pressure that is high enough to offset the crossover disruption is not easy. As a result, the selection pressure should be adaptively adjusted in response to the degree of difficulty of the problems without actually varying the tournament size.

PARAMETERS.

E_{chrom}: Elite chromosome, N_{chrom}: New chromosome

STEP 2*. Generate one chromosome from the probability vector
```
if the first generation then
    Echrom:=generate(p);    /*Initialize the elite chromosome*/
Echrom:=generate(p);        /*Generate a new chromosome*/
```
STEP 3*. Let them compete and let the winner inherit persistently
```
winner,loser:=compete(Echrom,Nchrom);
Echrom:=winner;             /*Update the elite chromosome*/
```

Fig. 4.2. Modification of the cGA that realizes the pe-cGA.

Since pairwise tournament selection has been employed, the selection pressure should be adaptively increased by evolving only two competing chromosomes. It has already been shown that the selection pressure is also increased by passing the best chromosome(s) onto the next generation (i.e., elitism in Sect. 4.2.2). Therefore, the idea is to increase the selection pressure in accordance with the difficulty of the problems by employing elitism in an judicious manner. In order to accomplish this, STEPS 2–3 of the cGA (in Fig. 4.1) should be replaced by the ones described in Fig. 4.2.

The procedures are being added with a view to simulating elitism in the cGA. Of the two competing chromosomes, only the loser is replaced by the new one that is generated from the PV. In other words, the winner is never eliminated in so far as a better chromosome has not yet been produced from the PV. This scheme is called *persistent elitist compact GA* (pe-cGA).

The following theorem is important in this regard.

Theorem 4.1: The pe-cGA is equivalent to the (1+1)-Evolutionary Strategy (ES) with self-adaptive mutation.

Proof: Let $g : R^m \to R$ be the objective function to be maximized. Consider the Markovian process $\mathbf{X} = \{\mathbf{X}_k; k \geq 0\}$ generated by the stochastic algorithm [98]

$$\mathbf{X}_{k+1} = \begin{cases} \mathbf{X}_k + l_k \mathbf{Z}_k, & \text{if } g(\mathbf{X}_k + l_k \mathbf{Z}_k) > g(\mathbf{X}_k) \\ \mathbf{X}_k, & \text{otherwise} \end{cases} \tag{4.1}$$

where l_k is the step length control parameter that is increased as long as mutation improves solutions. Each random vector $\mathbf{Z}_k$ (of the sequence of independent and identically distributed random vectors) has a joint probability density function (pdf) with independent marginal densities.

The model of Eq. (4.1) falls within the purview of the (1+1)-ES with self-adaptive mutation, if the step length control parameter is changed when the relative frequency of improving mutations is below or above some threshold within τ trials [98]. In other words, Eq. (4.1) can exactly model the (1+1)-ES with self-adaptive mutation if it considers the τ trials as an elementary event in stage k.

Now, let $\mathbf{Y}_k$ be a random vector generated from the PV. The probability distribution of $\mathbf{Y}_k$ is given by

$$F_{\mathbf{Y}_k}(y_1, \cdots, y_m) = \prod_{i=0}^{m} \mathbf{P}_k(i) \qquad (4.2)$$

where m is the number of decision variables and $\mathbf{P}_k(i)$ represents the probability distribution of the ith decision variable in the kth generation. Moreover, m is given by n/ν where n and ν are the length of chromosome and the number of bits used for encoding a decision variable, respectively.

With $\mathbf{Y}_k$, the pe-cGA takes the form

$$\mathbf{X}_{k+1} = \begin{cases} \mathbf{Y}_k, & \text{if } g(\mathbf{Y}_k) > g(\mathbf{X}_k) \\ \mathbf{X}_k, & \text{otherwise.} \end{cases} \qquad (4.3)$$

Here, $\mathbf{Y}_k$ describes a new chromosome generated from the PV. Also, $\mathbf{X}_k$ represents a winner (i.e., the elite chromosome) in the $(k-1)$th generation, viz., a chromosome in the kth generation that is inherited from $(k-1)$-th generation. Then, Eq. (4.3) can naturally be rewritten as

$$\mathbf{X}_{k+1} = \begin{cases} \mathbf{X}_k + (\mathbf{Y}_k - \mathbf{X}_k), & \text{if } g(\mathbf{Y}_k) > g(\mathbf{X}_k) \\ \mathbf{X}_k, & \text{otherwise.} \end{cases} \qquad (4.4)$$

The random vector $\mathbf{Z}_k$ and its scaling constant l_k in Eq. (4.1) have been used. Since $\mathbf{Y}_k$ is also a random vector and $\mathbf{X}_k$ is a constant vector in the kth generation, one can relate $\mathbf{Z}_k$ with $\mathbf{Y}_k$ as follows:

$$l_k \mathbf{Z}_k = \mathbf{Y}_k - \mathbf{X}_k. \qquad (4.5a)$$

Here, the pdf of $\mathbf{Z}_k$ can be computed easily by

$$f_{\mathbf{Z}_k}(z_1, \cdots, z_m) = f_{\mathbf{Z}_k}(\mathbf{z}) = |l_k| f_{\mathbf{Z}_k}(l_k \mathbf{z} + \mathbf{X}_k). \qquad (4.5b)$$

Employing Eqs. (4.4) and (4.5), Eq. (4.3) can be rewritten as

$$\mathbf{X}_{k+1} = \begin{cases} \mathbf{X}_k + l_k \mathbf{Z}_k, & \text{if } g(\mathbf{X}_k + l_k \mathbf{Z}_k) > g(\mathbf{X}_k) \\ \mathbf{X}_k, & \text{otherwise.} \end{cases} \qquad (4.6)$$

Comparing Eqs. (4.1) and (4.6), one can conclude that the two algorithms (i.e., pe-cGA and (1+1)-ES with self-adaptive mutation) follow an identical model. ∎

The general tendency is to treat EDAs and ESs as belonging to different realms of evolutionary algorithms. *Theorem 4.1* is interesting in that it reduces an EDA with elitism to a certain ES. Thus, the pe-cGA not only achieves a higher performance but also opens up an avenue for examining the unsuspected relationship between EDAs and ESs.

Instead of gaining in selection pressure, the pe-cGA may lose genetic diversity owing to inherent elitism. From *Theorem 4.1*, however, we see that the pe-cGA maintains genetic diversity that is comparable with that of (1+1)-ES with self-adaptive mutation. Furthermore, the pe-cGA offers some additional benefits over (1+1)-ES to GA practitioners and designers. First, the step length control parameter l_k is adjusted in a dynamic manner. Second, the pe-cGA does not have to select and fix the multivariate probability distribution for generating the random vector $\mathbf{Z}_k$. In other words, the mutation distribution of pe-cGA can also be adaptively adjusted.

In the (1+1)-ES with self-adaptive mutation, however, the update rule for l_k should be effective and the probability distribution for $\mathbf{Z}_k$ should be selected with care. This is because they directly and critically affect the performance of the algorithm. Popular choices for mutation distribution are Gaussian and Cauchy distributions [98, 119].

Lee and Yao [66] developed an EA using stable Lévy distribution with different values of parameters in this regard. The objective is to adaptively alter the mutation distribution in the environment. However, the Lévy parameter distribution is not self-adaptive and the adjusted mechanism is applied only to the variation of the decision variables, not to the self-adaptive deviation (i.e., l_k).

4.3.2 Nonpersistent Elitist Compact Genetic Algorithm

It may be noted that strong elitism may lead to premature convergence (to a suboptimal solution). This is because a high selection pressure brought about by strong elitism results in the population's reaching equilibrium very fast, but it inevitably sacrifices genetic diversity. Thus, a parameter η that indicates an allowable scope of the elite chromosome's inheritance is introduced to control the strength of elitism. This parameter restrains the scope of inheritance (i.e., the number of generations) of the winner, thereby playing a role in retrieving genetic diversity to some extent. This scheme is called *nonpersistent elitist compact GA* (ne-cGA). In order to employ elitism in a nonpersistent manner, the STEPS 2–3 (in Fig. 4.1) should be modified by the procedures described in Fig. 4.3.

As in pe-cGA, the loser is always replaced by a new candidate generated from the current PV. However, the winner can be passed on to the next generation only when the present depth of inheritance, denoted by θ, does not exceed the allowable scope of inheritance (i.e., η). In other words, a new randomly generated chromosome can replace the winner if $\theta > \eta$. Physically, this is very similar to mutation or random immigrant mechanism in GAs. Hence,

```
PARAMETERS.
   Echrom: Elite chromosome, θ: Current depth of inheritance
   Nchrom: New chromosome, η: Allowable scope of inheritance

STEP 2**. Generate one chromosome from the probability vector
   if the first generation then
       θ:=0                        /*Initialize the control parameter*/
       Echrom:=generate(p);   /*Initialize the elite chromosome*/
   Echrom:=generate(p);       /*Generate a new chromosome*/
STEP 3**. Let them compete and let the winner inherit nonpersistently
   winner,loser:=compete(Echrom,Nchrom);
   if θ ≤ η and winner==Echrom then
       θ++;              /*Increment the control parameter*/
   else if winner≠ Echrom then
       Echrom:=winner; /*Update the elite chromosome by the winner*/
       θ:=0;              /*Reset the control parameter*/
   else
   /*Replace the winner as a new chromosome randomly generated*/
       Echrom:=generate(p[i]:=0.5,∀i)
       θ:=0;              /*Reset the control parameter*/
```

Fig. 4.3. Modification of the cGA that realizes the ne-cGA.

this strategy may gently nudge the simulated population towards restoration of genetic diversity. Moreover, the ne-cGA has almost all the characteristics of the pe-cGA because it also employs elitism (the strength is restricted, though).

The following theorem relates to the scope of inheritance η.

Theorem 4.2: The allowable scope of inheritance (η) should not exceed the simulated population size N. That is, $\eta < N$.

Proof: The letters $\mathbf{X}_k$ and $\mathbf{Y}_k$ are as in *Theorem 4.1*. Define $\mathbf{W}_k$ as a random vector generated from a random PV set to 0.5. Let $\mathbf{V}_k$ be another random vector defined as $(\mathbf{P}_{k+1} - \mathbf{P}_k)$. This vector represents the changes between intergeneration PVs. Let us assume that the winner takes the PV to convergence regardless of optimality of the solution. In other words, it is assumed that the (current) winner always defeats its competitor when the PV converges, irrespective of whether the solution is optimal or suboptimal.

The evolution of PV can be described by

$$\mathbf{P}_{k+1}(i) = \begin{cases} \mathbf{P}_k(i) + E[\mathbf{V}_k(i)|\mathbf{X}_k(i) = 1], & \text{if } \mathbf{X}_k(i) = 1 \\ \mathbf{P}_k(i) + E[\mathbf{V}_k(i)|\mathbf{X}_k(i) = 0], & \text{if } \mathbf{X}_k(i) = 0. \end{cases} \tag{4.7}$$

Each conditional expectation on $\mathbf{V}_k(i)$ can be computed as follows:

$$E[\mathbf{V}_k(i)|\mathbf{X}_k(i) = 1] = (1/N) \cdot p[\mathbf{V}_k(i) = 1/N|\mathbf{X}_k(i) = 1]$$

$$+ 0 \cdot p[\mathbf{V}_k(i) = 0 | \mathbf{X}_k(i) = 1] \tag{4.8a}$$

$$E[\mathbf{V}_k(i) | \mathbf{X}_k(i) = 0] = (-1/N) \cdot p[\mathbf{V}_k(i) = -1/N | \mathbf{X}_k(i) = 0]$$
$$+ 0 \cdot p[\mathbf{V}_k(i) = 0 | \mathbf{X}_k(i) = 0]. \tag{4.8b}$$

On the other hand, each conditional probability is seen to be

$$p[\mathbf{V}_k(i) = 1/N | \mathbf{X}_k(i) = 1] = p[\mathbf{Y}_k(i) = 0] = 1 - \mathbf{P}_k(i) \tag{4.9a}$$

$$p[\mathbf{V}_k(i) = 0 | \mathbf{X}_k(i) = 1] = p[\mathbf{Y}_k(i) = 1] = \mathbf{P}_k(i) \tag{4.9b}$$

$$p[\mathbf{V}_k(i) = -1/N | \mathbf{X}_k(i) = 0] = p[\mathbf{Y}_k(i) = 1] = \mathbf{P}_k(i) \tag{4.9c}$$

$$p[\mathbf{V}_k(i) = 0 | \mathbf{X}_k(i) = 0] = p[\mathbf{Y}_k(i) = 0] = 1 - \mathbf{P}_k(i). \tag{4.9d}$$

Consider Eq. (4.9a) first. To increase the ith element of PV by $1/N$ (i.e., $\mathbf{V}_k(i) = 1/N$) given that ith gene of the winner (i.e., $\mathbf{X}_k(i)$) has "1", the ith gene of its competitor (i.e., $\mathbf{Y}_k(i)$) should generate "0" because the winner in the $(k-1)$th generation (i.e., $\mathbf{X}_k$) becomes a winner in the present kth generation (i.e., the chromosome is passed to the next, viz., $(k+1)$th generation). Since $\mathbf{P}_k(i)$ is the probability that $\mathbf{Y}_k(i) = 1$, it is computed as $1 - \mathbf{P}_k(i)$. The rest of Eq. (4.9) can be explained in a similar manner.

Using Eqs. (4.8) and (4.9), Eq. (4.7) can be rewritten as

$$\mathbf{P}_{k+1}(i) = \begin{cases} \mathbf{P}_k(i) + (1/N) \cdot \{1 - \mathbf{P}_k(i)\}, & \text{if } \mathbf{X}_k(i) = 1 \\ \mathbf{P}_k(i) - (1/N) \cdot \mathbf{P}_k(i), & \text{if } \mathbf{X}_k(i) = 0. \end{cases} \tag{4.10}$$

Define $\mathbf{P}_k^M$ and $\mathbf{P}_k^m$ as $\max_{\forall j} \mathbf{P}_k(j)$ and $\min_{\forall j} \mathbf{P}_k(j)$, respectively. None of the elements in the PV should be brought to convergence by the same winner because there is no guarantee that the chromosome leads to an optimal solution. It means that $\mathbf{P}_k^M$ or $\mathbf{P}_k^m$ should not be taken to convergence to "1.0" or "0.0" by the same winner. Hence, follow inequalities (4.11a) and (4.11b):

$$\mathbf{P}_{k+\eta}^M < 1 \tag{4.11a}$$

$$\mathbf{P}_{k+\eta}^m > 0. \tag{4.11b}$$

Here, η is the allowable scope of inheritance of the winner.

By employing Eq. (4.10), Eq. (4.11a) can be rewritten as follows:

$$\mathbf{P}_{k+\eta}^M = (1 - 1/N) \cdot \mathbf{P}_{k+\eta-1}^M + 1/N$$
$$= (1 - 1/N)^2 \cdot \mathbf{P}_{k+\eta-2}^M + \{(1 - 1/N) + 1\} \cdot (1/N)$$
$$= \cdots$$
$$= (1 - 1/N)^\eta \cdot \mathbf{P}_k^M + \{1 - (1 - 1/N)^\eta\}$$
$$\text{(since the simulated population size } N \gg 1)$$
$$\approx (1 - \eta/N) \cdot \mathbf{P}_k^M + \eta/N < 1.$$

Thus, the allowable scope η is seen to satisfy

$$\eta < N. \tag{4.12a}$$

In a similar manner, Eq. (4.11b) can be rewritten as

$$\begin{aligned}
\mathbf{P}^m_{k+\eta} &= (1 - 1/N) \cdot \mathbf{P}^m_{k+\eta-1} \\
&= (1 - 1/N)^2 \cdot \mathbf{P}^m_{k+\eta-2} \\
&= \cdots \\
&= (1 - 1/N)^\eta \cdot \mathbf{P}^m_k \\
&\approx (1 - \eta/N) \cdot \mathbf{P}^m_k > 0.
\end{aligned}$$

The allowable scope η is also found to obey the inequality

$$\eta < N. \tag{4.12b}$$

By considering Eqs. (4.12a) and (4.12b), it is seen that the allowable scope η should not exceed the simulated population size N.   ■

It follows from *Theorem 4.2* that the quality of solution found by ne-cGA is identical to that found by pe-cGA when $\eta = N$. This is experimentally confirmed later (Sect. 4.5.5). It is obvious that the ne-cGA does not require any extra memory (as in the pe-cGA). Moreover, the ne-cGA is not demanding on computational costs either (see Fig. 4.3).

4.4 Speedup Model

This section presents a *speedup* model (i.e. a gain in convergence speed). Due to the similarity of stochastic mechanisms of sGA and cGA, the respective convergence speeds would not be very different (especially as population size increases). At the very least, the speed of cGA is slightly higher than that of sGA. Furthermore, the convergence speed of ne-cGA is approximately the same as that of pe-cGA from a statistical point of view because the ne-cGA imposes a somewhat relaxed elitism on the pe-cGA. Therefore, the speedup S is defined as the ratio of the number of fitness evaluations of cGA (i.e., T_{cGA}) to that of pe-cGA (i.e., T_{pe-cGA}).
That is

$$S = \frac{T_{cGA}}{T_{pe-cGA}}. \tag{4.13}$$

Consider a well-known convergence model of population based GAs in the context of problems in which BBs are of equal salience, genes converge uniformly, and the fitness is distributed binomially. One-max problem is a typical example. It is formulated as follows:

$$f_{OneMax} = \sum_{i=1}^m x_i \tag{4.14}$$

where m is the number of BBs (i.e., bits) and x_i is the value of the ith gene.

From the characteristics of the problem, we can approximate the mean and variance of fitness of the population as a normal distribution with mean μ_t, and variance σ_t^2. Here, $\mu_t = mp_t$ and $\sigma_t^2 = mp_t(1 - p_t)$, and p_t represents the proportion of correct BBs (of the population) in generation t. Mühlenbein and Schlierkamp-Voosen [73] proposed a convergence model for the problem and ordinal selection schemes as follows:

$$\mu_{t+1} = \mu_t + I\sigma_t. \tag{4.15}$$

Here, I is the selection intensity that is defined as the expected increase in the average fitness of a population after the selection operation.

Equation (4.15) leads to [73]

$$p_t = \frac{1}{2}\left[1 + \sin\left(\frac{I}{\sqrt{m}}t + \arcsin(2p_0 - 1)\right)\right]. \tag{4.16}$$

The convergence time t_{conv} (i.e., the number of generations encountered before convergence occurs) can be shown to be:

$$t_{conv} = \left(\frac{\pi}{2} - \arcsin(2p_0 - 1)\right)\frac{\sqrt{m}}{I}. \tag{4.17}$$

If the population size is N, the number of function evaluations (T_{conv}) performed before convergence is clearly given by

$$T_{conv} = \left(\frac{\pi}{2} - \arcsin(2p_0 - 1)\right)\frac{N\sqrt{m}}{I}. \tag{4.18}$$

From Eqs. (4.13) and (4.18), the speedup S is seen to be

$$S = \frac{T_{cGA}}{T_{pe-cGA}} = \frac{I_{pe-cGA}}{I_{cGA}}. \tag{4.19}$$

Hence, the speedup can also be computed as the ratio of selection intensity of pe-cGA to that of cGA. This is quite reasonable because the selection intensity is inversely proportional to the speed of convergence. Note that the selection intensity is required for computing the speedup. Unfortunately, it is not the same for different selection schemes. For a tournament selection of size s, for example, the selection intensity I is given by [10]

$$I = \mu_{s:s} = s\int_{-\infty}^{\infty} x\phi(x)(\Phi(x))^{s-1}dx. \tag{4.20}$$

Here, $\phi(x) = \exp(-x^2/2)/\sqrt{2\pi}$ and $\Phi(x) = \int_{-\infty}^{x}\phi(z)dz$ are the probability density function (pdf) and the cumulative distribution function (cdf) respectively of a standard normal distribution (with zero mean and unit standard deviation).

To find the selection intensity of cGA, let us consider the selection method that randomly chooses two individuals and reflects two copies of the better one on the population. This is equivalent to a steady-state pairwise tournament selection [46]. Since there is no population of selected parents for a steady-state scheme, the selection intensity can be redefined as the expected fitness increase after N offsprings have been generated [112]. In the case of selection of cGA with a tournament size s, the best individual competes with the other $(s-1)$ individuals, updating the PV along the way [46]. It means that all the parents in these $(s-1)$ competitions (i.e., $2(s-1)$ individuals) are replaced by the copies of the best individual out of s individuals. Thus, the creation of N offsprings is equivalent to the execution of $N/\{2(s-1)\}$ selections. Since one execution of the selection with tournament size s increases the expected fitness for population size N by $(1/N)\sum_{i=1}^{s-1}(\mu_{s:s}-\mu_{i:s})$, the selection intensity of cGA can be defined by

$$I_{cGA} = \frac{1}{2(s-1)}\sum_{i=1}^{s-1}(\mu_{s:s} - \mu_{i:s}) \tag{4.21}$$

where $\mu_{i:s}$ represents the expected fitness value of the ith ranked individual of a random sample of size s of a population.

The expected value of the ith-order statistic $\mu_{i:s}$ can be computed by [10, 112]

$$\mu_{i:s} = s\binom{s-1}{i-1}\int_{-\infty}^{\infty} x\phi(x)\Phi(x)^{i-1}(1-\Phi(x))^{s-i}dx. \tag{4.22}$$

When $s = 2$ (i.e., a steady-state pairwise tournament selection), the selection intensity is given as

$$
\begin{aligned}
I_{cGA} &= \frac{\mu_{2:2} - \mu_{1:2}}{2} \\
&= \int_{-\infty}^{\infty} x\phi(x)\Phi(x)dx - \int_{-\infty}^{\infty} x\phi(x)\{1 - \Phi(x)\}dx \\
&= \int_{-\infty}^{\infty} x\phi(x)\{2\Phi(x) - 1\}dx \\
&= 2\int_{-\infty}^{\infty} x\phi(x)\Phi(x)dx \approx 0.56.
\end{aligned}
\tag{4.23}
$$

It is interesting to note that the selection intensity for the tournament selection (of sGA) with $s = 2$ (see Eq. (4.20)) is exactly the same as that of cGA with $s = 2$, viz., Eq. (4.23). That is, convergence performances of sGA and cGA are identical when $s = 2$. However, it is observed that the convergence speed of sGA is slightly lower than that of cGA (see Fig. 4.4(b)). The discrepancy seems to be due to the inherent nature (i.e., a sort of hitchhiking) of the process of mixing of BBs.

On the other hand, the pe-cGA reflects two copies of a tournament winner on the population.[4] However, the winner is deterministically chosen as one of the competitors. In the context of a population, the selection mechanism can be approximated by truncation selection. The corresponding truncation selection takes the top $(1/5)$ of individuals in a population as parents because a selected individual whose fitness value (of normal distribution) is higher than the fitness of the top $(1/5)$ individuals would be a winner with *prob.* ≥ 0.98 from the long-run behavior point of view.

Moreover, truncation selection that picks the top $(1/\delta)$ portion of the population as parents is equivalent to a (μ, λ) selection with $\mu = \lambda/\delta$ [74]. Therefore, the selection mechanism of pe-cGA amounts to the (μ, λ) selection with $\mu = \lambda/5$ (i.e., $\delta = 5$). Since the selection intensity of (μ, λ) selection can be given by $(\lambda/\mu) \cdot \phi(\Phi^{-1}(1 - \mu/\lambda))$, the pe-cGA has the following intensity:

$$I_{pe-cGA} = 5 \cdot \phi(\Phi^{-1}(0.8)) \approx 1.46. \tag{4.24}$$

By substituting Eqs. (4.23) and (4.24) in Eq. (4.19), the speedup S is found to be

$$S = \frac{T_{cGA}}{T_{pe-cGA}} = \frac{I_{pe-cGA}}{I_{cGA}} = 2.61. \tag{4.25}$$

It is noted that the indicated assumption invalidates this model for the problems whose some genes (i.e., BBs) are highly correlated and are of unequal salience. However, such characteristics carry a tendency that prevents the cGA from steadily converging to a solution because of the absence of a strong reference for convergence. That is, the speedup plays a role in providing a lower bound on such problems (i.e., $S \geq 2.61$).

4.5 Experimental Results and Discussion

In this section, the performance of pe-cGA and ne-cGA are compared with that of cGA, sGA, and (1+1)-ES on various test functions and Ising Spin-Glasses (ISG) systems (a real-world application) through computer simulations. Fitness value (e.g., the number of correct BBs) and the number of (fitness) function evaluations are taken to be performance measures. The former considers solution quality (i.e., optimality) and the latter indicates the convergence performance.

In all the experiments, the sGA uses tournament selection without replacement and uniform crossover with exchange probability 0.5 [46]. The crossover is applied with probability one and the mutation probability is set to 0.0. The parameter η directly influencing the strength of elitism (i.e., selection pressure) is set to $0.1N$ as a default value. All the results were averaged over 100 runs. Each experiment is terminated when the PV converges to a solution.

---

[4] It is equivalent to the steady-state pairwise tournament selection of cGA.

Table 4.1. Statistical comparison of algorithms ($N = 100$) on f_{OneMax}.

Measure	sGA	cGA	pe-cGA	ne-cGA
μ_{conv}	3150.0	2836.4	1127.6	1142.1
σ_{conv}	82.27	88.11	86.72	69.09

Statistical t-test			
Measure	sGA − cGA	cGA − ne-cGA	ne-cGA − pe-cGA
t-value	26.046[†]	151.296[†]	1.813
Order	pe-cGA ∼ ne-cGA ≻ cGA ≻ sGA		

[†] The value of t is *significant* at $\alpha = 0.01$ by a paired, two-tailed test. The symbols ≻ and ∼ represent *dominance* and *indifference* between algorithms.

4.5.1 Results for the Problems Involving Lower Order BBs

A 100-bit one-max problem (i.e., the counting ones problem) and a minimum deceptive problem (mDP) formed by concatenating ten copies of minimum deceptive function [38] are considered for evaluating the proposed algorithms on problems involving lower-order BBs. The one-max problem and the mDP are representative problems with the order-one BBs and the order-two BBs, respectively.

A 100-bit one-max problem (i.e, f_{OneMax}) that is specified by Eq. (4.14) is considered first. Figures 4.4(a) and 4.4(b) compare the number of correct BBs (i.e., bits) and the number of function evaluations returned by each algorithm as applied to one-max problem. On the face of it, the proposed algorithms do not find high quality solutions while convergence speeds are admittedly far higher than those of sGA and cGA. Table 4.1 supports the claim on the improvement of convergence speed of the proposed algorithms over sGA and cGA. Moreover, the validity of the proposed speedup can also be concluded from Table 4.7.

In the interest of fair comparison of the algorithms on the basis of optimality and convergence performance, we investigate the number of correct BBs (i.e., solution quality) obtained by each population size that performs the same number of function evaluations [3], [22]. Since the population sizes of 35 (sGA), 40 (cGA), 90 (pe-cGA), and 100 (ne-cGA) perform approximately 1200 function evaluations (see Fig. 4.4(b)), for example, the solutions returned by the populations should be compared for investigating the superiority of the algorithms. Unfortunately, finding the exact population size for a particular execution for each GA is very difficult in practice. However, determining the population size with certain constraints is relatively easy. We can determine the population size for each GA by exhaustive search so as to achieve almost the same number of function evaluations. From this perspective, the number of correct BBs is plotted against function evaluations in Fig. 4.5. The figure

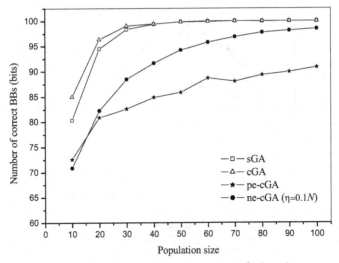

(a) Number of correct BBs versus population size.

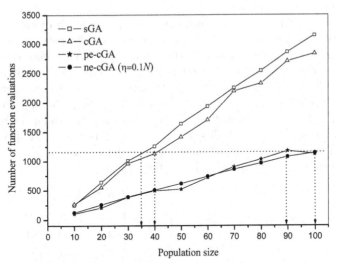

(b) Number of evaluations versus population size.

Fig. 4.4. Performance of algorithms on f_{OneMax}.

shows that the sGA, cGA, and ne-cGA achieve similar quality of solution while the pe-cGA computes a worse solution. The reason is discussed below.

The pe-cGA may suffer from lack of genetic diversity leading to premature convergence since an elite chromosome is never lost unless a superior chromosome appears. This is the reason why the ne-cGA restricts the scope

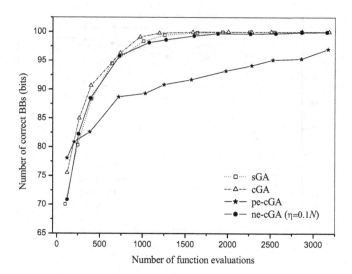

Fig. 4.5. Fair comparison of algorithms on f_{OneMax}.

of inheritance of the elite chromosome (i.e., the strength of elitism). In the figures, we may note that ne-cGA augments genetic diversity to some extent, thereby maintaining the quality of solution and the convergence speed at levels comparable with those of the reference algorithms. The pe-cGA, however, returns a poor overall performance. Note that the unsatisfactory performance of the pe-cGA may not be critical from a practical point of view because most real-world problems such as multicast routing, and (adaptive) equalizer design in fading channels, ISG systems, and maximum satisfiability problem etc., cannot be modeled as combinations of order-one BBs.

The next test problem is a mDP defined by

$$f_{mDP} = \sum_{i=1}^{m} f(x_{2i}) , \quad \text{where } f(x_{2i}) = \begin{cases} 0.7, & \text{if } x_{2i} = 00 \\ 0.4, & \text{if } x_{2i} = 01 \\ 0.0, & \text{if } x_{2i} = 10 \\ 1.0, & \text{otherwise.} \end{cases} \quad (4.26)$$

Here, x_{2i} presents the values (i.e., alleles) of a 2-bit long substring (i.e., BB).

Figures 4.6(a) and 4.6(b) depict the quality of solution and the speed of convergence of the algorithms when applied to the mDP with $m = 10$ (i.e., 10-BBs). Trends similar to those found in Figs. 4.4(a) and 4.4(b) can be seen here as well. Table 4.2 also shows the higher convergence speed of the proposed algorithms over sGA and cGA. Moreover, accuracy of the model of speedup can be observed in Table 4.7. With regard to fair comparison, Fig. 4.7 shows the quality of solution versus the number of function evaluations. In the figure, it is seen that ne-cGA outperforms cGA and performs as well as sGA.

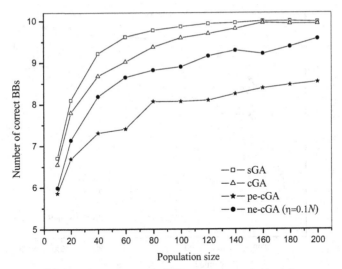

(a) Number of correct BBs versus population size.

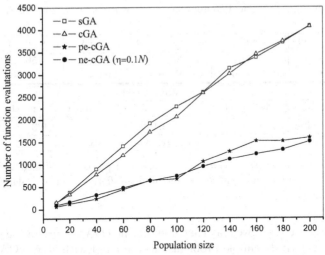

(b) Number of evaluations versus population size.

Fig. 4.6. Performance of algorithms on f_{mDP}.

However, there is a difference with regard to pe-cGA. That is, the pe-cGA may be less effective than sGA and cGA on real-world problems because some of them can be modeled as combinations of order-two BBs (especially, the deceptive ones).

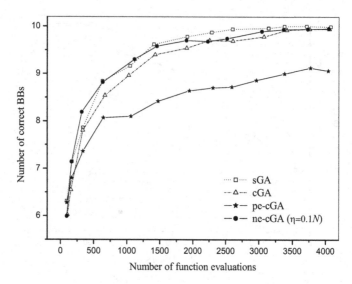

Fig. 4.7. Fair comparison of algorithms on f_{mDP}

Table 4.2. Statistical comparison of algorithms ($N = 200$) on f_{mDP}.

Measure	sGA	cGA	pe-cGA	ne-cGA
μ_{conv}	4076.0	4066.4	1531.3	1505.2
σ_{conv}	313.70	395.16	262.05	225.90

	Statistical t-test		
Measure	sGA $-$ cGA	cGA $-$ pe-cGA	pe-cGA $-$ ne-cGA
t-value	0.191	52.162[†]	2.463
Order	ne-cGA $\sim$ pe-cGA $\succ$ cGA $\sim$ sGA		

[†] The value of t is *significant* at $\alpha = 0.01$ by a paired, two-tailed test. The symbols $\succ$ and $\sim$ represent *dominance* and *indifference* between algorithms.

From Figs. 4.4–4.7, we may conclude that ne-cGA is a promising candidate for solving relatively simple problems as compared with sGA, cGA, and pe-cGA, even though a significant improvement of overall performance is not evident.

4.5.2 Results for the Problems Involving Higher Order BBs

Fully deceptive problems [27] are considered for testing pe-cGA and ne-cGA on the problems involving higher-order BBs. Before investigating the performance, we define a trap function f_{trap} (that is a constituent of deceptive problems [27]) by

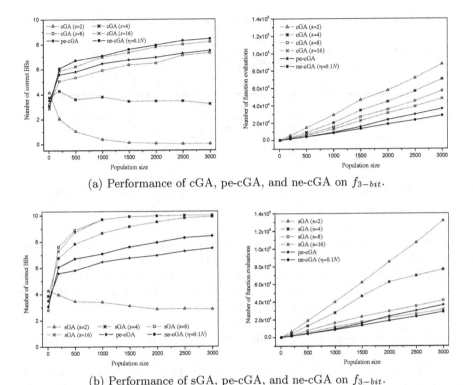

(a) Performance of cGA, pe-cGA, and ne-cGA on f_{3-bit}.

(b) Performance of sGA, pe-cGA, and ne-cGA on f_{3-bit}.

Fig. 4.8. Performance comparison of algorithms on f_{3-bit}.

$$f_{trap}(u, a, b, z, k) = \begin{cases} (a/z) \cdot (z - u), & \text{if } u \leq z \\ \{b/(k - u)\} \cdot (u - z), & \text{otherwise.} \end{cases} \qquad (4.27)$$

where u is the unitation that is defined as the number of ones of a (sub)string, a and b are the local (i.e., deceptive) and the global optimum respectively, z is the slope-change location, and k is the (sub)problem size.

The first deceptive problem is based on a three-bit trap function. The test problem is formed by concatenating ten copies of the three-bit trap function for a total chromosome length of 30 bits. Each three-bit trap function has a deceptive-to-optimal ratio of 0.7. That is, the problem is formulated by

$$f_{3-bit} = \sum_{i=1}^{10} f_{trap}(u_{3i}, 0.7, 1, 2, 3) \qquad (4.28)$$

where u_{3i} is the unitation of a 3-bit long substring.

Figure 4.8 depicts the results of each algorithm as applied to the first deceptive problem. From Fig. 4.8(a), it is observed that pe-cGA achieves a better solution (than does cGA) with $s = 8$ and ne-cGA finds a quality of

Table 4.3. Statistical comparison of algorithms ($s = 16$, $N = 3000$) on f_{3-bit}.

Measure	sGA	cGA	pe-cGA	ne-cGA
μ_{conv}	31050.0	48716.3	35896.73	28415.41
σ_{conv}	2872.3	4518.37	3765.04	2967.25

	Statistical t-test		
Measure	cGA − pe-cGA	pe-cGA − sGA	sGA − ne-cGA
t-value	$21.797^{\dagger}$	$10.234^{\dagger}$	$6.381^{\dagger}$
Order	ne-cGA $\succ$ sGA $\succ$ pe-cGA $\succ$ cGA		

† The value of t is *significant* at $\alpha = 0.01$ by a paired, two-tailed test. The symbols $\succ$ and $\sim$ represent *dominance* and *indifference* between algorithms.

solution that is comparable with that found by cGA with $s = 16$. At the same time, their convergence speeds are higher than those of all the cGAs. From Fig. 4.8(b), it is also clear that the quality of solution found by the proposed algorithms is no better than that found by the sGA with $s = 4$ while their convergence speeds are higher than those of sGAs except when $s = 16$ for pe-cGA. Convergence performance is clearly seen in the statistical test in Table 4.3. Moreover, Table 4.7 also supports the accuracy of the speedup model.

The reason why the proposed algorithms do not find a better solution than does the sGA with $s \geq 4$ is because they lack the memory to retain the knowledge about nonlinearity of the problems. However, this is an inherent characteristic of all the compact-type GAs. It is also seen that the solution quality returned by the proposed algorithms improves as the population size grows. Thus, if a larger population is employed, a higher quality of solution is obtained. Note the important fact that the algorithms (i.e., pe-cGA, ne-cGA) exert a selection pressure that is high enough to combat the disruptive effects of crossover without altering the tournament size. Furthermore, these benefits accrue without any compromise on memory requirements and computational costs. In the context of this test problem, on the other hand, it appears that the relation (investigated by Harik *et al.* [46]) between selection pressure and tournament size of the sGA and cGA is satisfied. The relation developed from a global schema theorem dictates the tournament size (of the sGA and cGA) that provides a higher selection pressure that is sufficient to grow the correct BBs as the population size increases. The tournament sizes in order-k BB should be greater than or equal to 2^{k-1} and 2^k, respectively, in the case of sGA and cGA [46].

The second deceptive problem is formed by concatenating ten copies of the four-bit trap function for a total chromosome length of 40 bits. Each four-bit trap function has a deceptive-to-optimal ratio of 0.7. That is, the problem is specified by

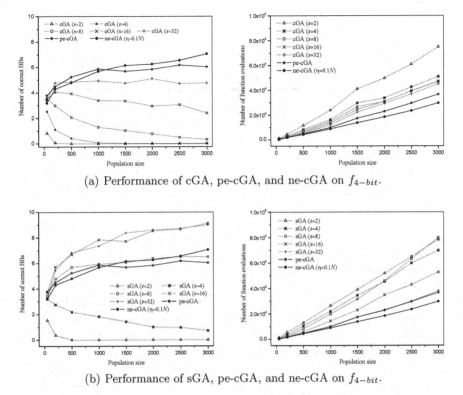

(a) Performance of cGA, pe-cGA, and ne-cGA on f_{4-bit}.

(b) Performance of sGA, pe-cGA, and ne-cGA on f_{4-bit}.

Fig. 4.9. Performance comparison of algorithms on f_{4-bit}.

$$f_{4-bit} = \sum_{i=1}^{10} f_{trap}(u_{4i}, 0.7, 1, 3, 4). \tag{4.29}$$

The results are compared in Fig. 4.9. As seen in Fig. 4.9(a), the proposed algorithms generally outperform all the cGAs that choose a tournament size s from 2 to 32. Moreover, the convergence speeds of pe-cGA and ne-cGA are higher than those of all the cGAs. In Fig. 4.9(b), it is seen that the proposed algorithms achieve a quality of solution that is similar to that obtained by sGA with $s = 8$ and their convergence speeds are higher than those of all the sGAs except when $s = 32$ for pe-cGA. The results are similar to those in Fig. 4.8. Fast convergence is supported by the statistical test of Table 4.4. Moreover, Table 4.7 also attests to the accuracy of the speedup model for the four-bit deceptive problem.

It is also clear that the relation suggested by Harik *et al.* [46] between selection pressure and tournament size is not satisfied for the cGA in this test problem. According to the relation, the correct BB will grow when the tournament size is greater than or equal to 16 (i.e., 2^k where $k = 4$). However, the cGA is not able to propagate the BB as long as the tournament size is less

Table 4.4. Statistical comparison of algorithms ($s = 32$, $N = 3000$) on f_{4-bit}.

Measure	sGA	cGA	pe-cGA	ne-cGA
μ_{conv}	37770.0	45014.17	36433.04	28737.64
σ_{conv}	5925.20	4711.80	4027.80	2435.51

Statistical t-test			
Measure	cGA − sGA	sGA − pe-cGA	pe-cGA − ne-cGA
t-value	9.570[†]	1.866	16.351[†]
Order	ne-cGA $\succ$ pe-cGA $\sim$ sGA $\succ$ cGA		

[†] The value of t is *significant* at $\alpha = 0.01$ by a paired, two-tailed test. The symbols $\succ$ and $\sim$ represent *dominance* and *indifference* between algorithms.

than 32 (i.e, $s < 32$) in this problem. It appears that there is some discrepancy in the relation. The tournament size may be intricately related to not only the deceptive-to-optimal ratio but also to the order of BBs and the number of collateral noise sources, etc., in the cGA. The rule [46] takes only the order of BBs into account. However, further investigation on these issues is beyond the scope of this investigation. Furthermore, such information may not be available in any case. It follows that the existing cGA may not be very useful in practice even if a perfect relation is discovered.

Figures 4.8 and 4.9 show that the proposed algorithms seem to adaptively adjust their selection pressures according to the difficulty of the problems. It is seen that the proposed algorithms are always able to provide a selection pressure that is enough to steadily grow the correct BBs as the population size increases. Therefore, they can effectively solve the difficult problems (especially, deceptive problems involving higher order BBs) without any knowledge about the problem dependent information such as the degree of deception, the order of BBs, and the selection pressure needed to combat disruptive effects of crossover.

4.5.3 Results for Continuous and Multimodal Problems

Most real-world problems do not involve simple concatenation of distinct order-k BBs since their solution/search spaces are continuous and multimodal. The problems can be modeled as an intricate combination of lower and higher order BBs. In order to investigate the performance on such problems, a circle function [77] and Schaffer's binary function [104] are employed. The functions may be used for modeling several real-world problems, especially those arising in the emerging areas of wireless networks such as the ultra-wide band antenna design and fading channel estimation problems.

The circle function to be minimized is investigated first. It is defined by

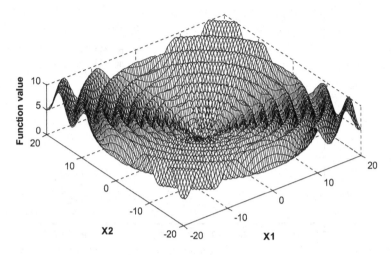

Fig. 4.10. Plot of two-dimensional f_C.

Table 4.5. Statistical comparison of algorithms ($N = 100$) on f_C ($n = 2$).

Measure	sGA	cGA	pe-cGA	ne-cGA
μ_{conv}	5047.0	4086.97	910.95	1174.3
σ_{conv}	833.21	600.65	119.22	153.89

Statistical t-test			
Measure	sGA − cGA	cGA − ne-cGA	ne-cGA − pe-cGA
t-value	2.338	58.586†	13.527†
Order	pe-cGA ≻ ne-cGA ≻ cGA ∼ sGA		

† The value of t is *significant* at $\alpha = 0.01$ by a paired, two-tailed test. The symbols ≻ and ∼ represent *dominance* and *indifference* between algorithms.

$$f_C(x) = \left(\sum_{i=1}^{n} x_i^2 \right)^{1/4} \cdot \left[\sin^2 \left(50 \left(\sum_{i=1}^{n} x_i^2 \right)^{1/10} \right) + 1.0 \right] \qquad (4.30)$$

where $x_i \in [-32.767, 32.768], \forall i$. Its two-dimensional landscape is plotted in Fig. 4.10. This multimodal function has many local optima (i.e., minima) that are located on concentric circles near the global optimum (i.e., the origin) [77].

Figures 4.11(a) and 4.11(b) compare the objective function values and the function evaluations achieved by the algorithms as applied to the circle function with $n = 2$. It is seen that the proposed algorithms significantly outperform sGA and cGA with regard to convergence speed (see Table 4.5), without unduly compromising on the solution quality. Under conditions of fair comparison as depicted in Fig. 4.12, the overall performance of ne-cGA

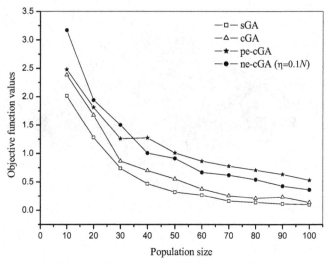

(a) Objective function values versus population size.

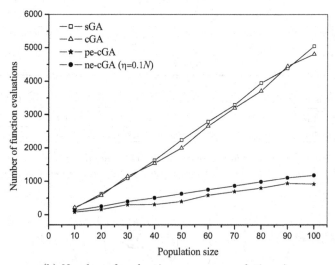

(b) Number of evaluations versus population size.

Fig. 4.11. Performance of algorithms on f_C with $n = 2$.

is better than that of the rest of the algorithms as the number of function evaluations increases while pe-cGA achieves a performance that is similar to that of sGA. As noted in Sect. 4.4, the model of speedup does not accurately estimate the extent of convergence improvement (see Table 4.7) due to inter-

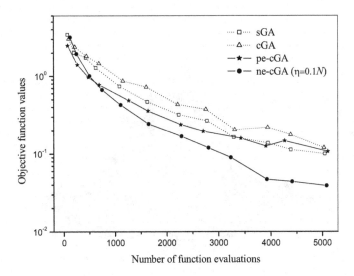

Fig. 4.12. Fair comparison of algorithms on f_C with $n = 2$.

dependency and unequal salience of BBs in this problem. Instead, it provides a marginal speedup for this kind of problem.

Schaffer's binary function to be maximized is considered next. The function is defined by

$$f_{S_6}(x) = \frac{\sin^2\left(\sqrt{\sum_{i=1}^{n} x_i^2}\right)}{1.0 + 10^{-3} \cdot \left(\sum_{i=1}^{n} x_i^2\right)^2} \tag{4.31}$$

where $x_i \in [-16.383, 16.384], \forall i$. The characteristics (e.g., landscape) of this function are easily grasped from its two-dimensional form shown in Fig. 4.13. The function is degenerate in the sense that many points share the same global optimal function value. As can be seen in Fig. 4.13, the points are located on the highest circle in the crown near the origin [104].

Figure 4.14 compares the algorithms as applied to Schaffer's binary function with $n = 5$. Figure 4.14(a) shows that the solution found by ne-cGA is better than those computed by sGA and cGA, while pe-cGA finds a solution that is similar to that returned by the cGA. Although the result is invalid when the population size is small, such populations are not regarded as feasible candidates in practice. Figure 4.14(b) and Table 4.6 show that their convergence speeds are higher than those of sGA and cGA. Furthermore, the improvement of convergence speed is about 8.38 times (see Table 4.7). It also supports the model of speedup as a lower bound. Accuracy is not high, however. Note that a larger discrepancy of speedup implies a higher inter-correlation and unequal salience of BBs.

On the basis of comparison studies, the proposed algorithms are found to be suitable for solving those types of problems. On the other hand,

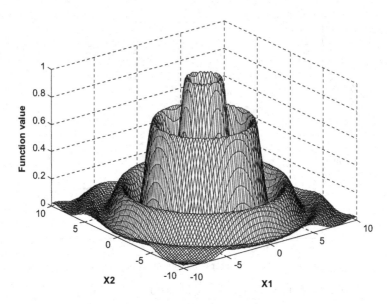

Fig. 4.13. Plot of two-dimensional f_{S_6}.

Table 4.6. Statistical comparison of algorithms ($N = 300$) on f_{S_6} ($n = 5$).

Measure	sGA	cGA	pe-cGA	ne-cGA
μ_{conv}	48798.0	42680.33	3721.83	5451.0
σ_{conv}	5406.12	3508.6	485.74	557.04
Statistical t-test				
Measure	sGA $-$ cGA	cGA $-$ ne-cGA	ne-cGA $-$ pe-cGA	
t-value	9.493†	104.794†	23.396†	
Order	pe-cGA $\succ$ ne-cGA $\succ$ cGA $\succ$ sGA			

† The value of t is *significant* at $\alpha = 0.01$ by a paired, two-tailed test. The symbols $\succ$ and $\sim$ represent *dominance* and *indifference* between algorithms.

Table 4.7. Comparison of speedup on test problems.

	Theory	f_{OneMax}	f_{mDP}	f_{3-bit}	f_{4-bit}	f_C	f_{S_6}
Speedup S	2.61	2.55	2.63	2.73	2.47	4.21	8.38

Figs. 4.11, 4.12 and 4.14 imply that sGA and cGA already have a selection pressure that is high enough to overcome crossover disruption because the degree of optimality improves as the population size increases.

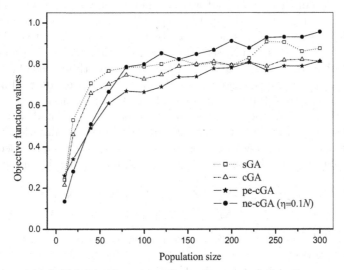

(a) Objective function values versus population size.

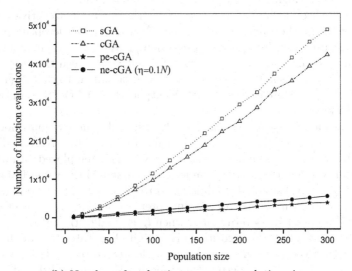

(b) Number of evaluations versus population size.

Fig. 4.14. Performance of algorithms on f_{S_6} with $n = 5$.

4.5.4 Comparison Results with Evolutionary Strategies

Evolutionary strategies (ESs) are among the main cutting-edge evolutionary algorithms. However, comparison of ESs with EDAs is quite unusual because they are considered to be different schemes. In other words, any criterion for comparing them in an unbiased way has not yet been suggested. This thesis has revealed a relationship between pe-cGA and (1+1)-ES in Sect. 4.3.1. More intense comparative study of the two schemes is imperative. Although ESs maintain at each step a set of solution candidates (i.e., a population), (1+1)-ES does not involve any idea of population. Without loss of generality, this work employs the average quality of solutions returned after (almost) the same execution time (i.e., the number of function evaluations) as the comparison criterion. All the test problems have been investigated. For discrete test problems, (1+1)-ES with Bernoulli distribution as a mutation distribution is taken as a reference. This is the same as (1+1)-EA proposed by He and Yao [49]. For an objective function g to be maximized, it can be defined as follows:

$$\mathbf{X}_{k+1} = \begin{cases} \mathbf{X}_k + \mathbf{Z}, & \text{if } g(\mathbf{X}_k + \mathbf{Z}) > g(\mathbf{X}_k) \\ \mathbf{X}_k, & \text{otherwise} \end{cases} \tag{4.32}$$

where $\mathbf{Z} = \{Z_1, \cdots, Z_n\}$, and Z_i is a Bernoulli random variable with a flipping probability of 0.1.

The solution quality for each algorithm is presented in Table 4.8. The (1+1)-ES is terminated at a specified epoch that is close to the convergence instant of all the algorithms. From Figs. 4.8 and 4.9, tournament sizes of sGA and cGA are seen to be $s = 4$ and $s = 8$ for the three-bit deceptive problem, respectively; and they are $s = 8$ and $s = 32$ for the four-bit deceptive problem. This is because the values provide a selection pressure that is enough to offset the disruptive effect of crossover. From Table 4.8, it is seen that the pe-cGA generally returns a solution that is better than or similar to that of (1+1)-ES except for f_{mDP}. Moreover, the ne-cGA outperforms (1+1)-ES on most of the test problems.

For continuous test problems, (1+1)-ES with self-adaptive mutation is employed as a reference. The (1+1)-ES is defined by Eq. (4.1). Normal distribution with zero mean and unit variance (i.e., $\mathcal{N}(0, 1)$) is employed for mutation. The initial step length control parameter l_0 is assigned the value 100 and the period τ (i.e., an instance for adjusting l_k) is set to 1. The step-length rule suggested in [98] with a view to avoiding premature convergence is employed. Evolution of (1+1)-ES is terminated when l_k is smaller than 0.1.

Performance of all the algorithms is exhibited in Table 4.9. In this table, it is seen that the pe-cGA is similar in performance to (1+1)-ES while the performance of ne-cGA is better. By observing the standard deviations, one can also conclude that (1+1)-ES is quite unstable with regard to solution quality as well as convergence speed. Note that the performance of (1+1)-ES would be the same as that of pe-cGA if the step length control parameter l_k and the mutation distribution $f_{\mathbf{Z}_k}(\mathbf{z})$ are adjusted to satisfy Eq. (4.5b) at every stage (*Theorem 4.1*). However, this is not possible in practice.

Table 4.8. Statistical comparison of algorithms on discrete test problems.

Problem	Measure	sGA	cGA	pe-cGA	ne-cGA	(1+1)-ES
f_{OneMax} (opt : 100)	μ_{QoS}	99.74	99.97	92.97	99.61	88.31
	σ_{QoS}	0.24	0.17	2.40	0.57	1.67
	μ_{conv}	1996.47	1988.93	1994.25	1988.54	2000
	σ_{conv}	73.2	50.98	228.97	64.58	-
f_{mDP} (opt : 10)	μ_{QoS}	9.76	9.71	8.56	9.73	9.80
	σ_{QoS}	0.50	0.48	1.10	0.51	0.50
	μ_{conv}	1951.95	1985.11	1919.41	1985.90	2000
	σ_{conv}	456.61	247.30	262.34	166.06	-
f_{3-bit} (opt : 10)	μ_{QoS}	7.51	5.58	6.74	7.26	6.93
	σ_{QoS}	0.98	1.39	1.45	1.24	1.39
	μ_{conv}	9930.40	9927.71	9659.63	9848.04	10000
	σ_{conv}	2520.62	543.70	878.89	1248.92	-
f_{4-bit} (opt : 10)	μ_{QoS}	5.97	5.02	6.08	7.05	5.04
	σ_{QoS}	1.17	1.42	1.32	1.05	1.32
	μ_{conv}	29876.4	29187.3	29583.1	29950.7	30000
	σ_{conv}	7556.91	2822.70	2838.26	2042.85	-

	Statistical t-test	
Problem	pe-cGA − (1+1)-ES	ne-cGA − (1+1)-ES
f_{OneMax}	15.919[†]	63.995[†]
f_{mDP}	−13.344[†]	−1.185
f_{3-bit}	−0.946	1.770
f_{4-bit}	5.375[†]	10.398[†]

	Statistical order	
Problem		
f_{OneMax}	pe-cGA ≻ (1+1)-ES	ne-cGA ≻ (1+1)-ES
f_{mDP}	(1+1)-ES ≻ pe-cGA	ne-cGA ∼ (1+1)-ES
f_{3-bit}	pe-cGA ∼ (1+1)-ES	ne-cGA ∼ (1+1)-ES
f_{4-bit}	pe-cGA ≻ (1+1)-ES	ne-cGA ≻ (1+1)-ES

[†] The value of t is *significant* at $\alpha = 0.01$ by a paired, two-tailed test. The symbols ≻ and ∼ represent *dominance* and *indifference* between algorithms.

As a consequence, the proposed algorithms (especially, ne-cGA) are also seen to be quite promising candidates for solving various problems[5] vis-à-vis the corresponding ESs.

4.5.5 Effects of the Scope of Inheritance

The ne-cGA is characterized by a parameter η that controls the inheritance scope of the elite chromosome. That is, it controls the strength of elitism.

---

[5] They may be approximated/modeled by the tested types of problems.

Table 4.9. Statistical comparison of algorithms on continuous test problems.

Problem	Measure	sGA	cGA	pe-cGA	ne-cGA	(1+1)-ES
f_C (opt : 0.0)	μ_{QoS}	0.2347	0.3051	0.1917	0.097	0.2788
	σ_{QoS}	0.2621	0.4104	0.1843	0.096	0.8766
	μ_{conv}	3016.14	3083.92	3020.27	3011.15	3094.22
	σ_{conv}	501.80	384.28	508.33	334.68	1496.71
f_{S_6} (0.994007)	μ_{QoS}	0.7180	0.6818	0.7790	0.8280	0.7711
	σ_{QoS}	0.2009	0.2059	0.1449	0.1703	0.2151
	μ_{conv}	2303.4	2341.5	2271.28	2340.20	2352.43
	σ_{conv}	340.47	257.01	187.89	256.12	1127.77

		Statistical t-test	
Problem		pe-cGA $-$ (1+1)-ES	ne-cGA $-$ (1+1)-ES
f_C		-0.553	-1.575
f_{S_6}		0.305	2.069†

		Statistical order	
Problem			
f_C		pe-cGA $\sim$ (1+1)-ES	ne-cGA $\sim$ (1+1)-ES
f_{S_6}		pe-cGA $\sim$ (1+1)-ES	ne-cGA $\succ$ (1+1)-ES

† The value of t is *significant* at $\alpha = 0.01$ by a paired, two-tailed test. The symbols $\succ$ and $\sim$ represent *dominance* and *indifference* between algorithms.

Hence, the parameter may affect the algorithm's performance. This experiment focuses only on examining the influence of this parameter on the performance. The algorithm is investigated on three test problems/functions: the one-max problem, the deceptive problem based on three-bit deceptive function, and Schaffer's binary function.

Figure 4.15 compares the number of correct BBs (i.e., bits) and the number of function evaluations returned by the pe-cGA and the ne-cGA with different values of η as applied to the one-max problem (i.e., a simple problem).

It is observed that the quality of solution improves as η decreases (i.e., elitism becomes weaker) except when the population size is small. This is due to increased genetic diversity brought about by weakening elitism. Moreover, this improvement does not have to pay any price in terms of reduced convergence speed (i.e., the number of function evaluations).

The solution quality of ne-cGA with $\eta = 0.05N$ is slightly better than that of ne-cGA with $\eta = 0.1N$ when the population size is larger than 60. However, their performances are nearly the same under fair comparison. Thus, the performance will not be improved any more when a value smaller than $0.05N$ is assigned to η. It implies that too low a value of elitism incurs degradation in algorithm's performance. This is the reason why the ne-cGA with $\eta = 0.05N$ cannot achieve a better performance with respect to optimality, as well as convergence speed when the population size is smaller than 60. We can see that elitism is quite weak when population is in the above range since

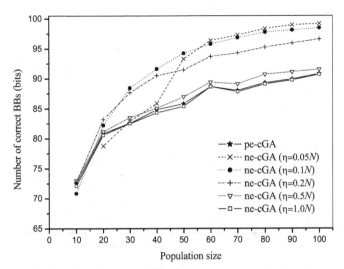

(a) Number of correct BBs (bits) versus population size.

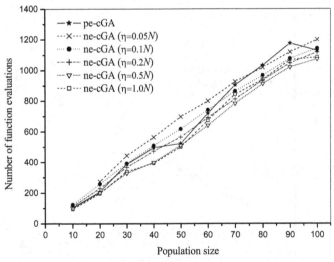

(b) Number of evaluations versus population size.

Fig. 4.15. Comparison of pe-cGA and ne-cGA with various η on f_{OneMax}.

the scope of inheritance is at most three generations. Thus, it is clear that a proper adjustment of elitism in one-max type problems leads to better quality (of solution) without unduly compromising on convergence performance.

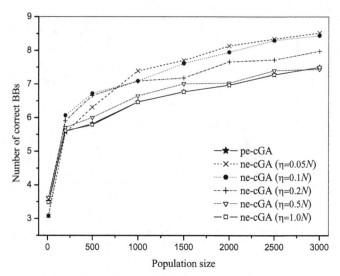

(a) Number of correct BBs versus population size.

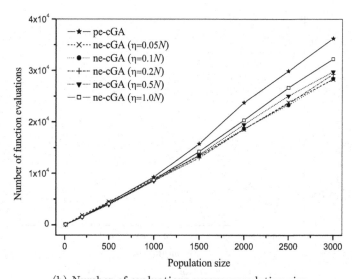

(b) Number of evaluations versus population size.

Fig. 4.16. Comparison of pe-cGA and ne-cGA with various η on f_{3-bit}.

Figure 4.16 compares the algorithms when applied to the deceptive problem constructed by a three-bit deceptive function with the deceptive-to-
optimal ratio of 0.7. The quality of solution improves as η decreases as long as
elitism is not too weak. Unlike the convergence tendency shown in Fig. 4.15,

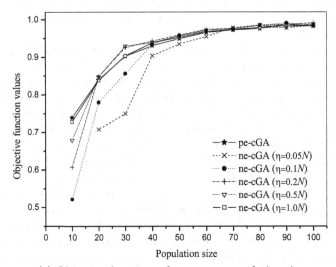

(a) Objective function values versus population size.

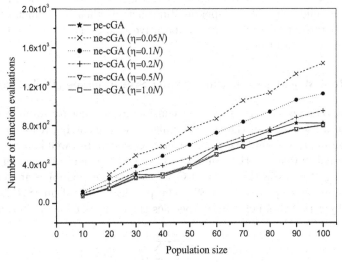

(b) Number of evaluations versus population size.

Fig. 4.17. Comparison of pe-cGA and ne-cGA with various η on f_{S_6} ($n = 5$).

however, the convergence performance also improves but slightly as long as a smaller value is assigned to η before it reaches $0.1N$.

Therefore, it is noted that a proper adjustment of elitism in deceptive-type problems (i.e., difficult problems) leads to a better solution without affecting the rate of convergence.

Figure 4.17 presents the effects of the parameter η on ne-cGA as applied to Schaffer's binary function with $n = 5$. In order to highlight the effects of restricted elitism, a 2-D function is considered. As can be seen in the figures, the convergence performance is getting worse without improved solution quality as a value less than $0.5N$ is assigned to the parameter. It is seen that the ne-cGA with $\eta = 0.5N$ returns the best performance when the solution quality and convergence speed are considered at the same time.

It follows that the reduction of η below half the simulated population size (i.e., $0.5N$) results in the degradation of both the solution quality as well as the convergence speed in this type of problems. However, this degradation is not of any major concern in so far as the parameter η is above $0.1N$.

As shown in Figs. 4.15–4.17, all the objective function values of ne-cGA with $\eta = 1.0N$ are comparable with those of the pe-cGA. Incidentally, it verifies an assertion that arises from *Theorem 4.2*. The assertion requires that the solution quality of ne-cGA with $\eta = N$ be the same as that of the pe-cGA.

As a result, it follows that an adequately controlled elitism[6] imparts genetic diversity, thereby improving the performance. Note that the adjustment of elitism would be problem-dependent in practice. Thus, it is difficult to properly characterize it. Based on the comparative studies, however, $\eta = 0.1N$ can be considered as a promising value.

4.5.6 Real-World Applications: Ising Spin-Glasses (ISG) Systems

Ising Spin-Glasses (ISG) systems are considered in order to examine the feasibility and usefulness of the proposed algorithms in solving real-world problems. Finding the ground state of a given ISG system is a well known problem in statistical physics. In the context of GAs, ISG systems are frequently employed as test cases in the study of GAs because they exhibit symmetry and a large number of plateaus [55, 89]. The physical state of an ISG system is defined by a Hamiltonian H that specifies the energy of the system by

$$H(\sigma) = -\sum_{i=0}^{n-1}\sum_{j=0}^{n-1} J_{ij}\sigma_i\sigma_j \qquad (4.33)$$

where a set of spins $\sigma = \{\sigma_0, \sigma_1, \cdots, \sigma_{n-1}\}$ taking values in $\{-1, +1\}$ represents the physical state of spins, and J_{ij} specifies a coupling coefficient from the ith spin to the jth spin. The objective is to find the state of spins so that the energy is minimized (i.e., the ground state is achieved).

---

[6] It denotes that the scope of the elite chromosome's inheritance is appropriately adjusted.

Table 4.10. Statistical comparison of algorithms on ISG systems.

Problem	Measure	sGA	cGA	pe-cGA	ne-cGA	(1+1)-ES
ISG	$\mu_{E/S}$	−1.2325	−1.2394	−1.3031	−1.3504	−1.2883
25-spins	$\sigma_{E/S}$	0.1956	0.1155	0.1298	0.0794	0.1371
ISG	$\mu_{E/S}$	−1.1334	−1.1530	−1.1029	−1.2116	−1.0748
100-spins	$\sigma_{E/S}$	0.0637	0.0819	0.0616	0.0423	0.0556

	Statistical t-test			
ISG	sGA − cGA	cGA − ES	ES − pe-cGA	pe-cGA − ne-cGA
25-spins	0.270	2.730†	0.783	3.110†
ISG	ES − pe-cGA	pe-cGA − sGA	sGA − cGA	cGA − ne-cGA
100-spins	3.392†	3.263†	1.890	5.923†

Problem	Statistical order
25-spins	ne-cGA $\succ$ pe-cGA $\sim$ (1+1)-ES $\succ$ cGA $\sim$ sGA
100-spins	ne-cGA $\succ$ cGA $\sim$ sGA $\succ$ pe-cGA $\succ$ (1+1)-ES

† The value of t is *significant* at $\alpha = 0.01$ by a paired, two-tailed test. The symbols $\succ$ and $\sim$ represent *dominance* and *indifference* between algorithms.

In our investigations, spins were arranged on a 2-D grid. The spins interact with only their nearest neighbors. We assume periodic boundary conditions. There exist several polynomial time algorithms for solving this special case of ISG problems. The optimal solution of such a system is provided by an online server [58]. Moreover, J_{ij} is chosen from $\{-1, +1\}$ in a random fashion (i.e., uniform distribution). All the algorithms were tested on two ISG systems with 25 and 100 spins arranged on 5×5 and 10×10 toroids, respectively.

Figure 4.18 and Table 4.10 exhibit the energy (per spin) found by each algorithm, measured after almost equal number of function evaluations. That is, the results of ISG systems with 25 and 100 spins are collected after approximately 10^4 and $3 * 10^4$ evaluations, respectively. It is observed that the ne-cGA finds the smallest energy for both systems although the pe-cGA can not find a solution that is better than that of sGA and cGA for the 100-spins system.

Based on the comparative studies presented in this section, we may conclude that the proposed algorithms, especially ne-cGA, are quite promising candidates for solving various types of problems – ranging from simple to difficult – including real-world applications.

4.6 Summary

This chapter has presented two elitism-based cGAs in an EDA framework. The aim is to address the problems associated with inadequate memory in the cGA by employing elitism in a proactive manner. The cGA is likely to

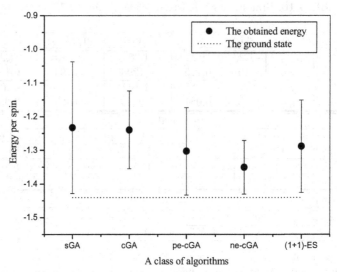

(a) Energy per spin for an ISG system with 25 spins.

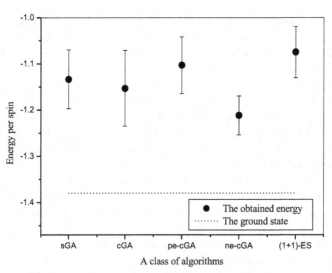

(b) Energy per spin for an ISG system with 100 spins.

Fig. 4.18. Performance of algorithms on ISG systems.

lose – irretrievably – the best current solution due to lack of memory. It cannot, therefore, speedily solve many difficult optimization problems in an efficient manner. The pe-cGA and the ne-cGA have been proposed to address this issue. The pe-cGA compensates for the lack of memory of the cGA by keeping the current best solution until, hopefully a better solution is found. It was shown that the pe-cGA is identical to (1+1)-ES with self-adaptive mutation. The ne-cGA relaxes selection pressure (i.e., elitism) of the pe-cGA by restricting the scope of elite chromosome's inheritance, thereby mitigating the possibility of premature convergence (i.e., convergence to a local optimum). The allowable scope of inheritance was shown to be bounded by the simulated population size (i.e., $\eta < n$). Furthermore, an analytic model of speedup for quantifying convergence improvement has been proposed.

Simulation studies showed that the proposed algorithms generally provide a better solution and a higher convergence speed than those of the sGA and cGA. The overall performance was shown to be better than those of sGA, cGA, and (1+1)-ES, especially as the problem becomes more and more difficult. It was also noted that most real-world problems cannot be modeled as simple problems. The experiments have also shown that the pe-cGA and ne-cGA do not have to adjust the tournament size in an attempt to exert a selection pressure high enough to compensate for crossover disruption. The algorithms always offer a proper selection pressure as if the tournament size (i.e., selection pressure) is automatically regulated in accordance with the degree of difficulty of the problems. Moreover, the experiments have shown that the speedup model accurately estimates the convergence improvement if the problems involve equally salient and uncorrelated BBs. It provides a marginal speedup, otherwise.

On the other hand, it was demonstrated that the quality of the solution improves as the scope of inheritance is decreased, while the convergence performance depends on the problem, in general. However, a proper adjustment of elitism improves the solution quality and enhances the convergence speed. Furthermore, a feasibility of the proposed algorithms for real-world applications has been demonstrated by investigating their performance on ISG systems.

The proposed algorithms can search the solution space effectively and speedily without compromising on memory and computational requirements. Moreover, the search capability may not depend overly on the shape of the search/solution space (i.e., difficulty of problems).

5

Real-coded Bayesian Optimization Algorithm

This chapter describes a real-coded (i.e., continuous) estimation of distribution algorithm (EDA) that solves real-valued (i.e., numerical) optimization problems of bounded difficulty quickly, accurately, and reliably. This is the *real-coded Bayesian optimization algorithm* (rBOA). The objective is to bring the power of (discrete) BOA to bear upon the area of real-valued optimization. That is, the rBOA must properly decompose a problem and effectively perform *probabilistic building-block crossover* (PBBC) for real-valued multivariate data. In other words, a unique feature of rBOA is to learn complex dependencies of variables and make use of mixture models at the level of substructures. To begin with, a Bayesian factorization is performed. The resulting that contains linkage information is then utilized for finding implicit subproblems (i.e., substructures). Mixture models are employed for independently fitting each of these substructures. Subsequently, an independent substructure-wise sampling draws the offspring.

This chapter also presents the scalability analysis of rBOA on problems of bounded difficulty. The scalability is measured by the growth of the number of (fitness) evaluations with the size of the problem until the optimum is reached. The total number of evaluations is computed by multiplying the population size for learning a correct probabilistic model (i.e, population complexity) and the number of generations until convergence (i.e., convergence time complexity).

The chapter is organized as follows. Section 5.1 briefly reviews EDAs. Section 5.2 outlines rBOA. Section 5.3 suggests a learning strategy for probabilistic models. Section 5.4 presents a popular technique for model sampling. Section 5.5 analyzes the scalability of rBOA. Real-valued test problems are cited in Sect. 5.6. Experimental results are presented in Sect. 5.7. We conclude with a summary in Sect. 5.8.

Chang Wook Ahn: *Advances in Evolutionary Algorithms: Theory, Design and Practice*, Studies in Computational Intelligence (SCI) **18**, 85–124 (2006)
www.springerlink.com   © Springer-Verlag Berlin Heidelberg 2006

5.1 Estimation of Distribution Algorithms

Estimation of distribution algorithms (EDAs), also known as *probabilistic model building genetic algorithms* (PMBGAs), signal a paradigm shift in genetic and evolutionary computation research [64, 90]. Incorporating (automated) linkage learning techniques into a graphical probabilistic model, EDAs exploit a feasible probabilistic model built around superior solutions found thus far while efficiently traversing the search space [90]. EDAs iterate the three steps listed below, until some termination criterion is satisfied:

1. Select good candidates (i.e., solutions) from a population of solutions.[1]
2. Estimate the probability distribution from the selected individuals.
3. Generate new candidates (i.e., offspring) from the estimated distribution.

It must be noted that the third step uniquely characterizes EDAs as it replaces traditional recombination and mutation operators employed by simple genetic and evolutionary algorithms (sGEAs). Although the sGEAs (with well-designed mixing operator) and EDAs deal with solutions (i.e., individuals) in quite different ways, it has been theoretically shown and empirically observed that their performances are quite close to each other [64, 90]. Moreover, EDAs ensure an effective mixing and reproduction of building blocks (BBs) due to their ability to accurately capture the (BB) structure of a given problem, thereby solving GA-hard problems with a linear or sub-quadratic performance in terms of (fitness) function evaluations [6, 20, 82, 89, 90]. However, there is a trade-off between the accuracy of the estimated distribution and the efficiency of computation. For instance, a close and complicated model is recommended if the fitness function to be evaluated is computationally expensive.

A large number of EDAs have been proposed for discrete and real-valued (i.e., continuous) variables in this regard. Depending on how intricate and involved the probabilistic models are, they are divided into three categories: *no dependencies*, *pairwise dependencies*, and *multivariate dependencies* [90].

The first category (i.e., no dependencies) is by far the simplest. It assumes statistical independence of all the variables in the problem. All the algorithms in this category compute the probability distribution as a product of univariate distributions. Looked at from the operational point of view, they approximate the order-one behavior of sGEA with uniform crossover. Hence, they are quite effective on linear problems where the variables have no interaction (i.e., order-one BBs) [64, 90]. Population based incremental learning (PBIL) algorithm [13], compact genetic algorithm (cGA) [46], and univariate marginal distribution algorithm (UMDA) [75] for discrete variables, and stochastic hill-climbing with learning by vectors of normal distributions (SHCLVND) [97], real-coded PBIL [106], and real-coded UMDA [61] for continuous variables are

---

[1] The initial population is randomly generated.

widely known in this respect. However, the assumption appears to be unrealistic in view of the fact that difficult optimization problems involve interactions. The independence assumption brings forth the disruption of superior partial solutions (i.e., BBs).

Several algorithms have been developed for countering the disruption of BBs by uniform crossover. Quantum-inspired evolutionary algorithm (QEA) [43], reinforcement learning estimation of distribution algorithm (RELEDA) [86], and persistent/nonpersistent elitist compact genetic algorithm (pe/ne-cGA) [4] are representative examples. Although they exhibit improved performance, they do not offer any solution to the basic problem.

As a first attempt in this direction, the category of pairwise dependencies has been studied. The assumption is that there are interactions only between pairs of variables. It is possible to quickly estimate the joint probability distribution by reflecting upon interactions between pairs of variables. Breeding and thoroughly mixing BBs of order-two, all the algorithms belonging to this category can efficiently solve linear as well as quadratic problems. Examples of this category include mutual information maximization for input clustering (MIMIC) [26], combining optimizers with mutual information tree (COMIT) [14], and bivariate marginal distribution algorithm (BMDA) [87] for discrete variables, and real-coded MIMIC [62] for continuous variables. However, difficult problems with higher-order interactions are beyond the reach of these algorithms.

The category of multivariate dependencies endeavors to use general probabilistic models, thereby solving many difficult problems quickly, accurately, and reliably [64, 90]. The more complex the probabilistic model the harder as well is the task of finding the best structure. At the expense of some computational efficiency with regard to learning the probabilistic model, they can significantly improve the overall time complexity for large (additively) decomposable problems due to their innate ability to reduce the number of computationally expensive fitness function evaluations. Extended compact genetic algorithm (ecGA) [44], factorized distribution algorithm (FDA) [76], estimation of Bayesian networks algorithm (EBNA) [61], and (hierarchical) Bayesian optimization algorithm ((h)BOA) [88, 89] are some leading examples for discrete variables.

Note that the BOA is perceived to be an important effort that employs general probabilistic models for discrete variables [6, 90]. It utilizes techniques for modeling multivariate data by Bayesian networks so as to estimate the (joint) probability distribution of promising solutions. A Bayesian network encodes the conditional probability of each variable given its parents. A simple greedy algorithm joining forces with the Bayesian Dirichlet equivalence(BDe) metric is used for constructing the Bayesian network. The BOA is very effective even on large decomposable (discrete) problems with loose and tight linkage of BBs. It is important to note that the power of BOA arises from realizing *probabilistic building-block crossover* (PBBC) that approximates *population-wise building-block crossover* by a probability distribution estimated on the basis of proper (problem) decomposition [88, 89]. The underlying decomposition can be performed regardless of types of dependency between variables because it is capable of accurately modeling any type of dependency due to the inherent characteristic (i.e., finite cardinality) of the discrete world. The PBBC can shuffle as many superior partial solutions (i.e., BBs) as possible in order to bring about an efficient and reliable search for the optimum.[2] Therefore, it is only natural that the principles of BOA be tried on real-valued variables.

In general, there are two approaches with regard to real-valued variables – *indirect* and *direct*. In order to effectively deal with real parameters, the former employs transform methods such as discretization and the latter estimates the parameters of predefined (mixture) distributions. BOA with discretization [92] follows the indirect approach, while estimation of Gaussian networks algorithm (EGNA) [62, 63], iterative density-estimation evolutionary algorithms (IDEAs) [17, 20], and mixed Bayesian optimization algorithm (MBOA) [81, 82] take the direct approach. A review of indirect and direct methods is presented in the sequel.

The BOA with discretization combines the strength of BOA for discrete representation with the advantage of evolutionary strategies (ESs) for real-valued (i.e., continuous) representation. In other words, the advanced recombination technique of BOA, viz., PBBC, is incorporated into the advanced mutation technique, viz., self-adaptive mutation for optimization of real-valued problems. Discretization is employed to transform solutions between the discrete and real-valued domains. However, it incurs computational overheads for discretizing the promising candidates for BOA and backing the discrete solutions into the continuous domain for adaptive ES. Furthermore, it is not scalable in the EDA framework since the complexity of required discrete probabilistic model exponentially increases with the target precision of the solution [82].

In the EGNA, the Gaussian network is induced in each generation by means of a chosen scoring metric (e.g., edge exclusion tests, Bayesian Gaussian equivalence (BGe), and Bayesian information criteria (BIC)) and the offspring is created by simulating the learned network. A simple greedy algorithm reveals good structures when the BGe and BIC scores are employed. However, the EGNA is not suitable for solving complicated problems because it only constructs a (simple) single-peak Gaussian model.

The IDEAs exploit Bayesian factorizations and mixture distributions for learning probabilistic models using the BIC just as the *expectation-maximization* (EM) or simple search algorithms do. There is a general, but simple factorization mixture selection to be named "mixed IDEA" (mIDEA) in this chapter. It clusters the selected individuals and subsequently estimates a factorized probability distribution in each cluster separately. It is evident that the mIDEA can learn various types of dependency. This is in contrast to the usage of a single model for the entire search space. However, it cannot

---

[2] That is, the maximum BB-wise mixing rate can be achieved.

realize the PBBC because different clusters that may create important BBs do not share all the common features.

The MBOA learns a Bayesian network with local structures in the form of decision trees coming with univariate normal-kernel leaves, thereby capturing the mutual dependencies of the variables. The main purpose of MBOA is to find a decomposition of the search space into subspaces, in which the parameters are mutually independent. The decomposition is encoded by the Bayesian network with decision trees and each resulting partition is locally modeled by the normal-kernel distribution. In the MBOA, one decision tree is built for each target variable, and the split nodes of the decision tree are used to linearly split the domain of parent variables into parts. The leaves represent the elementary models for obtaining the target variable. The MBOA builds the decision trees by recursively adding the split nodes for each variable until the BDe score returns a negative value for all the variables. This results in a decomposition of the conditional distribution's domain into axis-parallel partitions, thereby efficiently approximating the variables by univariate (kernel) distributions [83]. Although the MBOA can be very effective for problems involving variables with simple interactions (i.e., linearity), it is inefficient for nonlinear, symmetric problems because finding the linear split boundaries for detecting the inherent characteristics is very difficult and quite often even impossible.

It may be noted that the direct approach occupies a predominant position because the indirect approach fails to scale with problem size and solution precision [82].

5.2 Real-coded Bayesian Optimization Algorithm

This section describes the rBOA as an efficient tool for solving real-valued problems of bounded difficulty with a sub-quadratic scale-up behavior. The purpose is to transplant the strong points of BOA into the continuous world.

Generously drawing on generic procedures of EDAs (see Sect. 5.1), the following pseudo-code summarizes the rBOA:

STEP 1. INITIALIZATION
   Randomly generate initial population $\mathcal{P}$
STEP 2. SELECTION
   Select a set of promising candidates $\mathcal{S}$ from $\mathcal{P}$
STEP 3. LEARNING
   Learn a probabilistic model $\mathcal{M}$ from $\mathcal{S}$ using a metric (and constraints)
STEP 4. SAMPLING
   Generate a set of offspring $\mathcal{O}$ from the estimated probability distribution
STEP 5. REPLACEMENT
   Create a new population $\mathcal{P}$ by replacing some individuals of $\mathcal{P}$ with $\mathcal{O}$
STEP 6. TERMINATION
   If the termination criteria are not satisfied, go to STEP 2

In spite of similar behavior patterns, EDAs can be characterized by the method of learning a probabilistic model (in the STEP 3). That is, the performance of EDAs depends rather directly on the efficiency of probabilistic model learning. In general, the learning of probabilistic models consists of two tasks: learning the structure and learning the parameters [89], also known as *model selection* and *model fitting*, respectively [20]. The former determines the structure of a probabilistic model. The structure defines conditional dependencies and independencies. Model fitting estimates the (conditional) probability distributions with regard to the found structure.

It is noted that model selection is closely related to model fitting. In the model selection phase, the best structure is searched by investigating the values of a chosen metric for all possible structures. However, the results of model fitting are directly or indirectly needed for computing the metric. Due to the large number of possible structures, the outcome may be unacceptably high computational complexity unless model fitting is performed in some simple way. A detailed investigation is described in Sect. 5.3.1.

On the other hand, there is a significant difference between discrete and real-coded EDAs from the viewpoint of probabilistic model learning. Discrete EDAs can easily estimate a probability distribution for a given/observed data by simply counting the number of instances for possible combinations. Moreover, the estimated distribution converges to its true distribution as the data size increases. Thus, discrete EDAs can quickly and accurately carry out model selection and model fitting at the same time.

A typical attempt to bringing the merit of discrete EDAs to bear on real-valued variables is to use histogram methods. This follows from the observation that constructing the histogram for a discrete distribution from population statistics and approximating it for a continuous distribution are analogous tasks [114]. Of course, the problem is tricky in higher dimensions (as described in Sect. 5.1), but nonetheless, it is theoretically possible. Indeed, it converges as the population size tends to infinity.

On the other hand, real-coded EDAs cannot use the simple counting methods of discrete EDAs to estimate a probability distribution for real-valued data due to (uncountably) infinite cardinality. There is an efficient method for reliably approximating the true probability distribution. The method relies on *(finite) mixture models* [70]. Some recent methods for unsupervised learning of mixture models are capable of automatically selecting the exact number of mixture components and overcoming some drawbacks of the EM algorithm [35, 70]. Due to its iterative nature, however, reconciling the unsupervised mixture learning techniques with the EDA framework is obviously hopeless (regardless of the frequency of its use). In this regard, faster mixture models are believed to be useful for efficiently estimating the probability distribution, in spite of sacrificing the accuracy. Although the fast alternatives can significantly reduce the computational cost, they may not be suitable candidates as model fitting is required for every considered structure.

It is, therefore, impossible to directly employ the learning procedure of discrete EDAs (such as BOA) in order to learn a probabilistic model for real-valued variables. An alternative technique for learning probabilistic models in real space is needed. Such a technique can draw on the power of EDAs in the discrete domain. By incorporating the solution with offspring generation procedure (i.e., model sampling), the proper decomposition and the PBBC that are important characteristics of BOA can be realized. The solution is explained in Sect. 5.3.

5.3 Learning of Probabilistic Models

This section presents an efficient technique for learning probabilistic models. Two tasks stand out in this regard: model selection and model fitting.

5.3.1 Model Selection

Factorizations (or *factorized probability distributions*) discover dependencies and independencies among random variables. A factorization is a probability distribution that can be described as a product of *generalized probability density functions* (gpdfs) which are themselves *probability density functions* (pdfs) involving real-valued random variables [20, 30]. *Bayesian factorizations*, also known as *Bayesian factorized probability distributions* come under a general class of factorizations [20, 65]. A Bayesian factorization estimates a joint gpdf for multivariate (dependent) variables as a product of univariate conditional gpdf of each random variable. The Bayesian factorization is represented by a directed acyclic graph, called a Bayesian factorization graph, in which nodes (or vertices) and arcs identify the corresponding variables in the data set and the conditional dependencies between variables, respectively [20, 65].

An n-dimensional real-valued optimization problem is considered for discussion. We denote the random variables in the problem by $\mathbf{Y} = (Y_1, \cdots, Y_n)$ and their instantiations by $\mathbf{y} = (y_1, \cdots, y_n)$. The pdf of $\mathbf{Y}$ is represented by $f(\mathbf{Y})(\mathbf{y})$.[3]

In general, a pdf is represented by a probabilistic model $\mathcal{M}$ that consists of a structure ζ and an associated vector of parameters $\boldsymbol{\theta}$ (i.e., $\mathcal{M} = (\zeta, \boldsymbol{\theta})$) [17, 20]. As the rBOA employs Bayesian factorization, the pdf $f(\mathbf{Y})$ for the problem can be encoded as

$$f(\mathbf{Y}) = f_{(\zeta, \boldsymbol{\theta})}(\mathbf{Y}) = \prod_{i=1}^{n} f_{\hat{\boldsymbol{\theta}}^{Y_i}}(Y_i | \mathbf{\Pi}_{Y_i}) \tag{5.1}$$

where $\mathbf{Y} = (Y_1, \cdots, Y_n)$ presents a vector of real-valued random variables, $\mathbf{\Pi}_{Y_i}$ is the set of parents of Y_i (i.e., the set of nodes from which there exists an

---

[3] The second parenthesis of probability distribution can be omitted for convenience if it causes no ambiguity.

arc to Y_i), and $f_{\dot{\theta}^{Y_i}}(Y_i|\mathbf{\Pi}_{Y_i})$ is the univariate conditional pdf of Y_i conditioned on $\mathbf{\Pi}_{Y_i}$ with its parameters $\dot{\theta}^{Y_i}$.

Although there are various methods for learning the structure of a probabilistic model (i.e., model selection), a widely used approach has two basic factors: a scoring metric and a search procedure [20, 89]. The scoring metric measures the quality of the structures of probabilistic models (i.e., Bayesian factorization graphs) and the search procedure efficiently traverses the space of feasible structures for finding the best one with regard to a given scoring metric.

Scoring Metric

A penalized maximum likelihood criterion known as the *Bayesian information criterion* (BIC) is employed as the scoring metric. Although any metric can be used, the reason for choosing the BIC is its empirically observed effectiveness in greedy estimation of factorized probability distributions [17, 64, 76]. Let $\mathcal{S}$ be the set of selected individuals, viz., $\mathcal{S} = (\mathbf{y}^1, \cdots, \mathbf{y}^{|\mathcal{S}|})$, where $|\mathcal{S}|$ is the number of the individuals. The BIC that assigns the structure ζ a score is formulated as follows [17, 20]:

$$
\begin{aligned}
BIC(\zeta) &= \ln\left(\prod_{j=1}^{|\mathcal{S}|} f_{(\zeta,\theta)}(\mathbf{Y})(\mathbf{y}^j)\right) - \lambda \ln(|\mathcal{S}|)|\theta| \\
&= \sum_{j=1}^{|\mathcal{S}|} \ln f_{(\zeta,\theta)}(\mathbf{Y})(\mathbf{y}^j) - \lambda \ln(|\mathcal{S}|)|\theta|.
\end{aligned}
\tag{5.2}
$$

Here, λ regularizes the extent of penalty and $|\theta|$ is the number of parameters of $f_{(\zeta,\theta)}(\mathbf{Y})$. Physically, the first and second terms represent the model accuracy and the model complexity, respectively.

Computing the BIC score for the structure ζ requires its parameters θ which fit the structure. However, the relations of cause and effect among them lead to unacceptably high computational complexity. This is because the number of possible structures to be tested/traversed increases exponentially with the problem size and the parameter fitting for the data set in real space is by no means a simple undertaking.

In short, the impracticality arises from the close relationship between model selection and model fitting. One way to cross the hurdle is to break the connection without obscuring their intrinsic objectives. An important feature of model selection is to acquire an a priori knowledge of the variables which are conditionally dependent regardless of linearity, nonlinearity, or symmetry. The reason is that the dependent type itself is learned (with probability distributions) by model fitting (see Sect. 5.3.2). Decoupling the connection can be achieved by computing the needed probability distributions for possible structures from a reference distribution. This is so because computing a marginal distribution with regard to an interesting substructure from a (reference) probability distribution fitted on the whole problem space is much simpler than directly estimating the exact probability distribution corresponding to the real-valued data set. EGNA and IDEAs are widely known in this respect. However, this can be hazardous in that it may fail to discover specific dependencies such as nonlinearity or symmetry.

In order to overcome the difficulty, multiple (probability) distributions are employed instead of one (i.e., mixture distribution), with a view to capturing the specific dependencies by a combination of piecewise linear interactions. In other words, the probability distributions used should lead to correct structures by capturing the dependency itself. We define the correct structure as the Bayesian factorization graph that encodes only the true or false interactions of the variables, regardless of the types of dependencies. Moreover, we learn one structure because it has been shown empirically that using one suitably constructed structure is sufficient to solve difficult problems [6, 81, 88].[4]

We employ mixture models for efficiently modeling the selected individuals by a mixture of probability distributions. With this in view, the BIC in Eq. (5.2) must be modified further.

As the pdf $f_{(\zeta,\theta)}(\mathbf{Y})$ can be described by a linear combination of a number of mixture components, Eq. (5.2) can be extended to

$$BIC\,(\zeta) = \sum_{j=1}^{|\boldsymbol{S}|} \ln \left(\sum_{i=1}^{K} \alpha_i f_{(\zeta,\theta^i)}(\mathbf{Y})(\mathbf{y}^j) \right) - \lambda \ln \left(|\boldsymbol{S}| \right) \sum_{i=1}^{K} \left| \theta^i \right| \qquad (5.3)$$

where K is the number of mixture components, $\alpha_1, \cdots, \alpha_K$ are the mixing probabilities satisfying $\alpha_i \geq 0$, $\forall i$, and $\sum_{i=1}^{K} \alpha_i = 1$, and θ^i is the set of parameters defined on the ith mixture component.

The observed-data vector (i.e., the selected individuals $\boldsymbol{S}$) can be viewed as being incomplete due to the unavailability of the associated component-label vectors, $\mathbf{w}^1, \cdots, \mathbf{w}^{|\boldsymbol{S}|}$ [35, 70]. Each label $\mathbf{w}^i$ is a K-dimensional binary vector and each element w^i_j is set to 0 or 1, depending on whether $\mathbf{y}^j$ did or did not arise from the ith mixture component. The component-label vectors are taken to be the realized values of the random vectors, $\mathbf{W}^1, \cdots, \mathbf{W}^{|\boldsymbol{S}|}$, in which it is assumed that they agree with an unconditional multinomial distribution [70]. That is, the probability distribution of the complete-data vector carries an appropriate distribution for the incomplete-data vector. Hence, Eq. (5.3) can be rewritten as

$$BIC\,(\zeta) = \sum_{i=1}^{K} \sum_{j=1}^{|\boldsymbol{S}|} w^i_j \left\{ \ln \alpha_i + \ln f_{(\zeta,\theta^i)}(\mathbf{Y})(\mathbf{y}^j) \right\} - \lambda \ln \left(|\boldsymbol{S}| \right) \sum_{i=1}^{K} \left| \theta^i \right|$$

---

[4] However, it can be naturally extended to multiple structures.

$$= \sum_{i=1}^{K} \ln \alpha_i \sum_{j=1}^{|\mathcal{S}|} w_j^i + \sum_{i=1}^{K} \sum_{j=1}^{|\mathcal{S}|} w_j^i \ln f_{(\zeta,\boldsymbol{\theta}^i)}(\mathbf{Y})(\mathbf{y}^j) - \lambda \ln (|\mathcal{S}|) \sum_{i=1}^{K} |\boldsymbol{\theta}^i| .$$

$$(5.4)$$

As the vectors $\mathbf{w}^1, \cdots, \mathbf{w}^{|\mathcal{S}|}$ can be simulated by the resulting mixture distribution, it is natural that $\sum_{j=1}^{|\mathcal{S}|} w_j^i$ coincides with the expected number of selected individuals drawn from the probability distribution $f_{(\zeta,\boldsymbol{\theta}^i)}(\mathbf{Y})$, denoted by $|\mathcal{S}_i|$. As the maximal log-likelihood is equivalent to the maximal negative entropy, $\sum_{j=1}^{|\mathcal{S}|} w_j^i \ln f_{(\zeta,\boldsymbol{\theta}^i)}(\mathbf{Y})(\mathbf{y}^j) = -|\mathcal{S}_i| h \left(f_{(\zeta,\boldsymbol{\theta}^i)}(\mathbf{Y}) \right)$ where $h \left(f_{(\zeta,\boldsymbol{\theta}^i)}(\mathbf{Y}) \right)$ represents the differential entropy of $f_{(\zeta,\boldsymbol{\theta}^i)}(\mathbf{Y})$. Moreover, the number of parameters for each distribution is the same (i.e., $|\boldsymbol{\theta}'| \equiv |\boldsymbol{\theta}^1| = \cdots = |\boldsymbol{\theta}^K|$) because the structure ζ is fixed for every distribution to be mixed. Thus, Eq. (5.4) is rewritten as

$$BIC (\zeta) = \sum_{i=1}^{K} |\mathcal{S}_i| \left\{ \ln \alpha_i - h \left(f_{(\zeta,\boldsymbol{\theta}^i)}(\mathbf{Y}) \right) \right\} - K\lambda \ln (|\mathcal{S}|) |\boldsymbol{\theta}'| . \qquad (5.5)$$

Since the terms $|\mathcal{S}_i|$ and $\ln \alpha_i$ are not affected by the structure ζ, Eq. (5.5) can be further reduced to

$$BIC (\zeta) = - \sum_{i=1}^{K} |\mathcal{S}_i| h \left(f_{(\zeta,\boldsymbol{\theta}^i)}(\mathbf{Y}) \right) - K\lambda \ln (|\mathcal{S}|) |\boldsymbol{\theta}'| . \qquad (5.6)$$

Thus, the BIC in Eq. (5.6) leads to a correct factorization even if there is some kind of nonlinearity and/or symmetry between variables.

Search Procedure

Learning the structure of a probabilistic model given a scoring metric is NP-complete [20,50,89]. However, most EDAs have successfully employed a greedy approach for searching a promising structure with a chosen metric. We employ the *incremental greedy algorithm*, a kind of greedy search algorithm [50]. Being one among many variants, this greedy algorithm starts with an empty graph with no arcs, and proceeds by incrementally adding an arc (such that no cycles are introduced) that maximally improves the metric until no further improvement is possible. The greedy algorithm is not guaranteed to discover an optimal structure in general because searching for the structure is an NP-complete problem. However, the resulting structure is good enough for encoding the most important interactions between variables of the problem [88,89].

5.3.2 Model Fitting

Note that the BOA models any type of dependency because it maintains all the conditional probabilities in the learned structure, without losing any

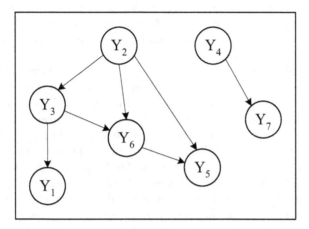

Fig. 5.1. Bayesian factorization graph involving component subproblems.

information due to finite cardinality of discrete variables. Moreover, the BOA naturally performs the PBBC with regard to the proper decomposition as it treats all the subproblems independently through the model selection, model fitting, and model sampling (i.e., offspring generation) phases. Hence, the BOA can solve difficult problems quickly, accurately, and reliably [88, 89].

With this in view, the model fitting phase (of the rBOA) must realise the probability distribution of a problem as a product of conditionally independent distributions accurately estimated on the basis of subproblems. In other words, the PBBC can be prepared by subspace-based model fitting. Unlike discrete EDAs, however, a preprocessing step for explicitly discovering subproblems (i.e., problem decomposition) is essential in real-coded EDAs, before performing the subspace-based model fitting. This is because discrete EDAs can implicitly carry out the problem decomposition in the course of (probabilistic) model learning while real-coded EDAs cannot do so (see Sect. 5.2).

Problem Decomposition

Problem decomposition can be easily accomplished because a set consisting of a node and its parents in the Bayesian factorization graph represents a component subproblem of decomposable problems. Here, the sets of variables of component subproblems may or may not be disjoint, but they cannot properly contain each other. In Fig. 5.1, the Bayesian factorization graph consists of five component subproblems, viz., $\{Y_2, Y_3\}, \{Y_3, Y_1\}, \{Y_2, Y_3, Y_6\}, \{Y_2, Y_6, Y_5\}, \{Y_4, Y_7\}$. However, it is not proper to directly use the component subproblems for model fitting. The reason is explained below.

The probability distribution of a problem can be constructed as a product of univariate conditional distributions which are computed from the probability distributions of component subproblems. Hence, the fitting process must be applied to every component subproblem. Since the fitting process itself is relatively complex even with a simple technique, it follows that fitting the model on the basis of component subproblems is not adequate, especially as the problem size increases.

Thus, an alternative decomposition is required for quickly and accurately performing model fitting on the basis of subproblems. In this regard, there is an observation that the set of a parent and its child nodes can be grouped as a kind of subproblem because the child nodes share a common feature even though they do not directly interact with each other. The set is called the *dual component* subproblem. It follows that the conditional distributions can be accurately computed from the probability distributions over the dual component subproblems. At this juncture, *minimal compound* subproblems are defined as the largest component or dual component subproblems that are not proper subsets of each other. In this way, a large number of fitting processes can be avoided (in proportion to the problem size) without losing fitting accuracy. For the problem in Fig. 5.1, the five component subproblems reduce to three minimal compound subproblems, viz., $\{Y_2, Y_3, Y_6, Y_5\}, \{Y_3, Y_1\}, \{Y_4, Y_7\}$ shown in Fig. 5.2(a).

There is another decomposition that is simple and also quite efficient for large problems. Consider the maximal connected subgraphs of a Bayesian factorization graph. Nodes in a maximally connected subgraph are looked on as a family; they have a common feature of being bound with common ancestors or descendants. Thus, the nodes can be thought of as interacting with each other in some sense. The conditional distributions can then be obtained from the probability distributions fitted over the maximally connected subgraphs without unduly compromising on the fitting accuracy. Here, the maximal connected subgraph is called the *maximal compound* subproblem. In Fig. 5.2(b), three minimal compound subproblems of Fig. 5.2(a) can be reduced to two maximal compound subproblems, viz., $\{Y_2, Y_3, Y_6, Y_5, Y_1\}, \{Y_4, Y_7\}$. Since this decomposition is a special case of decomposing the problem by minimal compound subproblems, minimal compound subproblems are employed for explaining the subspace-based model fitting.

Note that most real-coded EDAs in the category of multivariate dependencies (such as EGNA and IDEAs) choose an alternative that is far from being perfect. That is, conditional distributions are computed from the referencing distributions fitted over the problem space itself instead of subspaces. This cannot provide the PBBC, thereby resulting in an exponential scale-up performance. The reason is explained below.

BBs can be defined by groups of real-valued variables, each having values in some neighborhood (i.e., a small interval), that break up the problem into smaller chunks which can be intermixed to reach the optimum. Assume that the mixture models have been employed for model fitting. Univariate conditional distributions are computed from the mixture distributions fitted over the problem space itself. In the model sampling phase, an entire individual is drawn from a proportionately chosen mixture component. Regardless of the

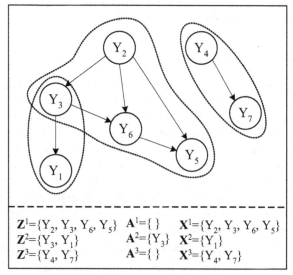

$$\mathbf{Z}^1=\{Y_2, Y_3, Y_6, Y_5\} \quad \mathbf{A}^1=\{\ \} \quad \mathbf{X}^1=\{Y_2, Y_3, Y_6, Y_5\}$$
$$\mathbf{Z}^2=\{Y_3, Y_1\} \qquad\quad \mathbf{A}^2=\{Y_3\} \quad \mathbf{X}^2=\{Y_1\}$$
$$\mathbf{Z}^3=\{Y_4, Y_7\} \qquad\quad \mathbf{A}^3=\{\ \} \quad \mathbf{X}^3=\{Y_4, Y_7\}$$

(a) Minimal compound subproblems.

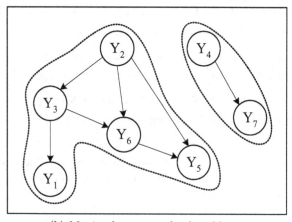

(b) Maximal compound subproblems.

Fig. 5.2. Examples of problem decomposition.

result of Bayesian factorization, it does not perform the PBBC as any mutual information of different regions cannot be shared. Instead, at least one mixture component must contain almost all the (superior) BBs of the problems for the sake of finding an optimal solution. In order to construct the mixture distribution that contain such mixture components, however, a huge population and a very large number of mixture components are required. It may result in an exponential scale-up behavior, even if the problem can be decomposable into subproblems of bounded order.

Subspace-based Model Fitting

Following proper decomposition, each substructure (corresponding to each subproblem) must be independently fitted. We employ the mixture models as an efficient tool for the purpose. The aim of mixture models is twofold: comprehending the type of dependency between variables and traversing the search space effectively. In general, higher-order factorized probability distributions are quite effective in discovering linear interactions between variables [20]. Each mixture component can model the linearity of the variables. Thus, the mixture models can approximate any type of dependency (e.g., nonlinearity or symmetry) by a combination of piecewise linear interaction models. In addition, it has the effect of partitioning each subspace for effective search.

Let $\mathbf{Z}^i = \left\{ Z_1^i, \cdots, Z_{|\mathbf{Z}^i|}^i \right\}$ be a vector of random variables of the ith subproblem in which the variables have already been topologically sorted for drawing new partial-individuals corresponding to the substructure. Note that the role of the topological sorting lies in discovering ancestral orders of the variables so as to generate the parents of each variables prior to the variable itself. Moreover, $\mathbf{Z}^i \not\subseteq \bigcup_{k=1}^{i-1} \mathbf{Z}^k$ and $\bigcup_i \mathbf{Z}^i = \mathbf{Y}$. Let $\mathbf{X}^i = \mathbf{Z}^i / \mathbf{A}^i$ where $\mathbf{A}^i = \mathbf{Z}^i \cap \left(\bigcup_{k=1}^{i-1} \mathbf{Z}^k \right)$. An example is given in Fig. 5.2(a).

Let $\zeta^{\mathbf{Z}^i}$ and $\boldsymbol{\theta}^{\mathbf{Z}^i}$ indicate a structure for the variables $\mathbf{Z}^i$ (i.e., substructure) and its associated parameters, respectively (viz., $\mathcal{M}^{\mathbf{Z}^i} = \left(\zeta^{\mathbf{Z}^i}, \boldsymbol{\theta}^{\mathbf{Z}^i} \right)$). Let $f_{\left(\zeta^{\mathbf{Z}^i}, \boldsymbol{\theta}^{\mathbf{Z}^i} \right)} \left(\mathbf{Z}^i \right)$ represent a pdf of $\mathbf{Z}^i$ and $f_{\left(\zeta^{\mathbf{A}^i}, \boldsymbol{\theta}^{\mathbf{A}^i} \right)} \left(\mathbf{A}^i \right) = \int_{\mathbf{X}_i} f_{\left(\zeta^{\mathbf{Z}^i}, \boldsymbol{\theta}^{\mathbf{Z}^i} \right)} \left(\mathbf{Z}^i \right) d\mathbf{X}_i$. As the mixture models are being employed, the pdf $f_{\left(\zeta^{\mathbf{Z}^i}, \boldsymbol{\theta}^{\mathbf{Z}^i} \right)} \left(\mathbf{Z}^i \right)$ can generally be represented by linearly combining $f_{\left(\zeta^{\mathbf{Z}^i}, \boldsymbol{\theta}_j^{\mathbf{Z}^i} \right)} \left(\mathbf{Z}^i \right)$ (for all j) that presents the pdf of the jth mixture component over $\mathbf{Z}^i$. Therefore, the pdf of $\mathbf{Y}$ can be written as a product of linear combinations of subspace-based (i.e., subproblem) pdfs as in Eq. (5.7),

$$f_{(\zeta, \boldsymbol{\theta})}(\mathbf{Y}) = \prod_{i=1}^{m} \sum_{j=1}^{c_i} \beta_{ij} \frac{f_{\left(\zeta^{\mathbf{Z}^i}, \boldsymbol{\theta}_j^{\mathbf{Z}^i} \right)} \left(\mathbf{Z}^i \right)}{f_{\left(\zeta^{\mathbf{A}^i}, \boldsymbol{\theta}_j^{\mathbf{A}^i} \right)} \left(\mathbf{A}^i \right)} \tag{5.7}$$

where m is the number of subproblems, c_i is the number of mixture components for $\mathbf{Z}^i$, β_{ij} is the mixture coefficient, $\beta_{ij} \geq 0$, and $\sum_{j=1}^{c_i} \beta_{ij} = 1$ for each i. In general, the mixture coefficient β_{ij} is proportional to the expected number of individuals of the jth mixture component of the subproblem $\mathbf{Z}^i$.

Any pdf can be rewritten as the product of univariate conditional pdfs according to its probabilistic model structure. Therefore, Eq. (5.7) can be rewritten as

$$f_{(\zeta,\theta)}(\mathbf{Y}) = \prod_{i=1}^{m} \sum_{j=1}^{c_i} \beta_{ij} \frac{\prod_{k=1}^{|\mathbf{Z}^i|} f_{\dot{\theta}_j^{Z_k^i}}\left(Z_k^i \big| \mathbf{\Pi}_{Z_k^i}\right)}{\prod_{l=1}^{|\mathbf{A}^i|} f_{\dot{\theta}_j^{A_l^i}}\left(A_l^i \big| \mathbf{\Pi}_{A_l^i}\right)}. \tag{5.8}$$

Therefore, the structure learned is efficiently fitted by the subspace-based mixture distributions even in the presence of nonlinearly and/or symmetrically dependent variables.

5.4 Sampling of Probabilistic Models

After model fitting, new individuals (i.e., offspring) are generated from sampling the resulting factorization, i.e., *model sampling*. With a view to generating the offspring, Eq. (5.8) can be simplified to

$$f_{(\zeta,\theta)}(\mathbf{Y}) = \prod_{i=1}^{m} \sum_{j=1}^{c_i} \beta_{ij} \prod_{k=1}^{|\mathbf{X}^i|} f_{\ddot{\theta}_j^{X_k^i}}\left(X_k^i \big| \mathbf{\Pi}_{X_k^i}\right). \tag{5.9}$$

Due to its simplicity and efficiency, the *probabilistic logic sampling* is employed [51]. Model sampling is performed in a straightforward manner. At first, the pdf of the jth mixture component for the ith subproblem is selected with a probability β_{ij}. Subsequently, a multivariate string (i.e., partial-individual) corresponding to $\mathbf{Z}^i$ can be drawn by simulating the univariate conditional pdfs of the chosen pdf which models one of the promising partitions (i.e., a superior BB) of a subspace (i.e., subproblem). By repeating this for all the subproblems, superior BBs can be mixed and bred for subsequent search.

To sum up, model selection amounts to a proper decomposition. The PBBC is realized successfully by model fitting and model sampling on the basis of the proper decomposition.

5.5 Scalability Analysis

This section analyzes the scalability of rBOA on problems of bounded difficulty, along the lines of approaches in [89,91].

5.5.1 Preliminaries

We make several assumptions for tractable analysis. First, we assume additively decomposable problems that consist of concatenated basis functions (i.e., subfunctions or subproblems); the overall fitness is the sum of contributions of all the subfunctions. It is also assumed that all the subfunctions are disjoint and their orders are all the same, viz., k. These are known as the *decomposable problems with bounded difficulty*.[5] Formally, the problems investigated in this analysis are characterized by

$$g(\mathbf{Y}) = \sum_{i=1}^{n/k} g_c \left(Y_{\boldsymbol{\omega}_{(i-1)k+1}}, \cdots, Y_{\boldsymbol{\omega}_{ik}} \right) \tag{5.10}$$

where k is the subproblem size, n is the problem size, $\boldsymbol{\omega}$ is a permutation of $\{1, \cdots, n\}$, and $g_c(\bullet)$ denotes a contribution of its arguments to the overall fitness.

Note that the assumption of equal order k may not be reasonable in practice. It is possible to set k to the average order of interactions for practical use [3]. Moreover, there is an upper bound on population complexity when k is set to the highest order of subproblems. No matter what methods are in use, there is no influence on the rBOA's scalability itself.

A particular partition (i.e., subproblem) is considered here. We denote the partition by a vector of random variables, $\mathbf{Z} = (Z_1, \cdots, Z_k)$, and its instantiation (i.e., subsolution) by a vector of real numbers, $\mathbf{z} = (z_1, \cdots, z_k)$. The individuals consisting of such partitions are also denoted by a vector of random variables, $\mathbf{Y} = (Y_1, \cdots, Y_k, \cdots, Y_n)$, where n is the size of the problem. The fitness of the individual $\mathbf{y}$ is indicated by $g(\mathbf{y})$. The total fitness of the individuals involving the subsolution (i.e., block) $\mathbf{z}$ is represented by a random variable $G(\mathbf{z})$. It is assumed that $G(\mathbf{z})$ follows a normal distribution, viz.,

$$G(\mathbf{z}) \sim \mathcal{N}(\mu_{G(\mathbf{z})}, \sigma_c^2) \tag{5.11}$$

where $\mu_{G(\mathbf{z})}$ is the average fitness of the individuals involving the block $\mathbf{z}$, and σ_c^2 is the total collateral noise variance coming from the rest of the variables. The collateral noise is defined as the disturbance in the decisions in regard to an observed partition. It is due to the fitness contributions of the rest of the partitions. The decision making process is explained in Sect. 5.5.2. Naturally, it follows that

$$\sigma_c^2 \propto n \tag{5.12}$$

where n is the problem size (i.e., the number of variables).

The normality assumption can be justified by the *central limit principle* in respect of all the decomposable problems consisting of uniformly scaled subproblems: the fitness contribution of each subproblem is of equal importance. The situation is the same even for exponentially scaled subproblems because the number of individuals required for building a good model does not increase and the number of generations until convergence grows linearly with the problem size [89, 91].

We also assume that decision variables can be modeled with normal mixture distributions. Given adequate data, the approximation to normal distribution is usually fairly close for some source components. In principle, the

---

[5] Many real-world problems can be approximated by decomposition into independent subproblems even when the subfunctions overlap [3, 89, 91].

assumption is also supported by the central limit theorem. In this regard, the rBOA employs normal mixture distributions in order to perform model selection (i.e., the BIC computation) and model fitting.

It is also assumed that the rBOA employs *truncation selection* for learning the probabilistic model due to its effectiveness over other selection methods in model learning. Recall that truncation selection with threshold τ ranks all the individuals according to their objective function values and selects the top τ-portion of the individuals. It is possible to carry out the analysis on other selection schemes. The theoretical results are expected to hold true for other selection methods [89, 91]. However, this is an issue for future research.

Note that the scalability of rBOA is measured by the growth of the number of (objective function) evaluations with regard to the problem size until reliable convergence to the optimum. Since the total number of evaluations E is computed as a product of the population size N and the number of generations T [89, 91], we have

$$E = \Theta(N \times T). \tag{5.13}$$

Therefore, the scalability analysis falls within the purview of investigating the population complexity and the convergence time complexity with respect to the size of the problem (i.e., the number of decision variables).

5.5.2 Population Complexity

In the process of discovering the problem regularities, viz., dependencies of variables, *decision* about *adding* and *not adding* an arc between two (real-valued) random variables, Z_1 and Z_2, in a particular partition $\mathbf{Z}$ has to be made [89, 91]. In order to decide whether an arc should be added or not between Z_1 and Z_2, the scores assigned by the BIC of Eq. (5.6) to the structures with and without arc are compared, and then the better alternative must be chosen. In the BIC, mixture distributions are incorporated to perform a correct factorization even in the presence of some nonlinearity and symmetry. Since the mixture model is composed of a linear combination of mixture components, we first investigate the BIC score with one component ($K = 1$), and then extend the result to multiple components. For notational convenience, $|\mathcal{S}|$ that denotes the selected population size in the BIC is replaced by N_s.

With regard to decision making, two possible cases are illustrated in Fig. 5.3. In the first case, Z_1 has no parents before deciding on whether or not to add the arc $Z_2 \rightarrow Z_1$ (Fig. 5.3(a)). In the second case, Z_1 already has a number of parents before the decision regarding the arc $Z_2 \rightarrow Z_1$ is considered (Fig. 5.3(b)). Taking into account the term involving Z_1, it is acceptable to compute a (selected) population size required for discovering the dependencies of variables because the BIC is decomposable. This is defined as the *critical (selected) population size* [89, 91]. The first case is analyzed below. The result is then extended to the second case.

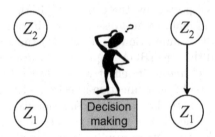

(a) Decision making when Z_1 has no parents.

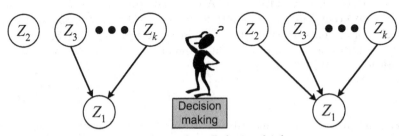

(b) Decision making when Z_1 has multiple parents.

Fig. 5.3. Examples on choosing between adding and not adding an arc.

The BIC score assigned to Z_1 without the arc $Z_2 \to Z_1$ is given by

$$BIC(Z_1) = -h(Z_1)N_s - 2\lambda \ln(N_s) \qquad (5.14)$$

where $h(Z_1)$ is the differential entropy of Z_1, and N_s is the number of selected solutions (i.e., selected population size). After adding the arc $Z_2 \to Z_1$, the BIC value of Z_1 is obtained by

$$BIC(Z_1|Z_2) = -N_s h(Z_1|Z_2) - 3\lambda \ln(N_s) \qquad (5.15)$$

where $h(Z_1|Z_2)$ is the conditional differential entropy of Z_1 given Z_2. To successfully discover the dependency between Z_1 and Z_2 (i.e., add the arc $Z_2 \to Z_1$), $BIC(Z_1 \leftarrow Z_2)$ should be greater than $BIC(Z_1)$. That is,

$$BIC(Z_1|Z_2) > BIC(Z_1). \qquad (5.16)$$

Substituting Eqs. (5.14) and (5.15) into Eq. (5.16) yields the following inequality:

$$(h(Z_1) - h(Z_1|Z_2))\, N_s - \lambda \ln(N_s) > 0. \qquad (5.17)$$

By replacing the mutual information term, $h(Z_1) - h(Z_1|Z_2)$, with a parameter I, Eq. (5.17) can be rewritten as

$$N_s - \frac{\lambda}{I} \ln(N_s) > 0. \qquad (5.18)$$

To compute the critical population size, denoted by N_s^{crit}, we must solve the following equation:

$$N_s^{crit} - \frac{\lambda}{I} \ln(N_s^{crit}) = 0. \qquad (5.19)$$

In Eq. (5.19), although two possible solutions exist when the ratio λ/I is sufficiently large, the larger solution is assigned to N_s^{crit} [89, 91]. Moreover, the mutual information of the normal random variables Z_1 and Z_2 is widely known to be

$$I = h(Z_1) - h(Z_1|Z_2) = \frac{1}{2} \ln \left(\frac{1}{1 - \rho_{Z_1,Z_2}^2} \right) \qquad (5.20)$$

where ρ_{Z_1,Z_2} is the correlation coefficient of Z_1 and Z_2, and $0 \leq |\rho_{Z_1,Z_2}| \leq 1$.
By employing Eq. (5.20), Eq. (5.19) can be rewritten as follows:

$$N_s^{crit} - \frac{2\lambda}{\ln \left(\frac{1}{1-\rho_{Z_1,Z_2}^2} \right)} \ln(N_s^{crit}) = 0. \qquad (5.21)$$

Let us consider a high $|\rho_{Z_1,Z_2}|$ (e.g., greater than 0.4). It represents strong dependency of Z_1 and Z_2. In this case, Eq. (5.21) gives a (very) small N_s^{crit} (less than 10), but such a small value is of no interest because it is not regarded as a feasible candidate in practice. However, it does not mean that the results of scalability analysis would not be valid for the strong dependencies of variables. Physically, the small N_s^{crit} for high $|\rho_{Z_1,Z_2}|$ means that strong dependencies can always be captured regardless of the size of population. Thus, $|\rho_{Z_1,Z_2}|$ has to be small enough to have a proper solution, N_s^{crit}. However, too small a value (e.g., less than 0.1) is of no interest since the learning process employing the solution N_s^{crit} might misjudge independent variables to be dependent.

Here is an important observation. The initial behavior of rBOA is crucial for successfully solving the problems. In the beginning, $|\rho_{Z_1,Z_2}|$ has a small value no matter how strongly dependent the variables Z_1 and Z_2 are in reality. This is because all the selected individuals are initially evenly distributed over the search space. In the first generation, the value of $|\rho_{Z_1,Z_2}|$ is the smallest. Once the individuals start crowding toward the optimum, $|\rho_{Z_1,Z_2}|$ would gradually increase so that the correct dependency can always be captured by the current size of the population. However, this might not come about if the dependency is not found at an early stage. Therefore, the scalability analysis focuses on the first generation of rBOA. This somewhat conservative approach is supported by empirical evidence (Sect. 5.7.3)

Note that the solution of Eq. (5.21) follows a power-law when $\ln(1/(1 - \rho_{Z_1,Z_2}^2))$ is small enough (i.e., λ/I is large enough in Eq. (5.19)). It implies that the solution N_s^{crit} takes the form $\alpha \left(1/\ln \left(\frac{1}{1-\rho_{Z_1,Z_2}^2} \right) \right)^{\beta}$, where α and β are some positive constant [89]. To approximate Eq. (5.21), the regulation

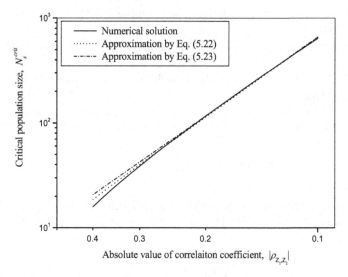

Fig. 5.4. Critical population size with regard to correlation coefficient.

parameter λ (of BIC) needs to be fixed in advance. Although λ is set to 0.5 in this study, any other value would do as well. The reason is that the constant term λ does not have any effect on the order of growth of N_s^{crit} (i.e., population complexity).

Therefore, we can approximate Eq. (5.21) by

$$N_s^{crit} \approx 2.1 \left(1 \Big/ \ln \left(\frac{1}{1 - \rho_{Z_1,Z_2}^2} \right) \right)^{1.25}. \tag{5.22}$$

where the constants $\alpha = 2.1$ and $\beta = 1.25$ are numerically obtained. The accuracy of the approximation for $0.1 \leq |\rho_{Z_1,Z_2}| \leq 0.4$ is shown in Fig. 5.4. The range is perceived to be promising for investigating the population complexity. In the figure, it is observed that the approximation is in close agreement with the numerical solution.

Since we are interested in the first generation of the rBOA, ρ_{Z_1,Z_2}^2 tends to be a small number (see Sect. 5.5.1). Thus, Eq. (5.22) can be further approximated as follows:

$$N_s^{crit} \approx 2.1 \left(\frac{1}{\rho_{Z_1,Z_2}^2} \right)^{1.25}. \tag{5.23}$$

The equation relates the population complexity to the correlation coefficient of variables. It apparently hints at the existence of some positive constants c_1 and c_2 such that $c_1 \cdot \frac{1}{\rho_{Z_1,Z_2}^2} \leq N_s^{crit} \leq c_2 \cdot \frac{1}{\rho_{Z_1,Z_2}^2}$ for any sufficiently large $\frac{1}{\rho_{Z_1,Z_2}^2}$. Thus, we obtain

$$N_s^{crit} = \Theta \left(\left(\frac{1}{\rho_{Z_1,Z_2}^2} \right)^{1.25} \right). \tag{5.24}$$

To analyze the scalability of rBOA, the growth of the solution N_s^{crit} must be investigated with regard to the problem size (i.e., the number of decision variables) [89, 91]. The relation between the correlation coefficient and the problem size is investigated in the sequel.

Note that the correlation coefficient of the two random variables Z_1 and Z_2 is given by $\rho_{Z_1 Z_2} = \frac{E[Z_1 Z_2] - E[Z_1]E[Z_2]}{\sigma_{Z_1}\sigma_{Z_2}}$. Consider two random vectors, $\mathbf{U} = (U_1, U_2)^T$ and $\mathbf{V} = (V_1, V_2)^T$, in order to investigate the relationship between the correlation coefficient and the moment. Assume that they have equal mean and variance, viz., $E[U_1] = E[V_1]$, $E[U_2] = E[V_2]$, $\sigma_{U_1}^2 = \sigma_{V_1}^2$ and $\sigma_{U_2}^2 = \sigma_{V_2}^2$. Although their first- and second-order moments are equal, the correlation coefficients are different if their second-order joint moments are not the same, viz., $E[U_1 U_2] \neq E[V_1 V_2]$. That is, the second-order joint moment retains, in essence, the attribute of the correlation coefficient of two variables. It indicates that the correlation coefficient ρ_{Z_1,Z_2} is directly connected with the (joint) probability distribution of Z_1 and Z_2. Therefore, we reach the following relation:

$$\rho_{Z_1,Z_2} \propto E[Z_1 Z_2]. \tag{5.25}$$

In the truncation selection with threshold τ, the top τ-portion of the population is selected as parents. The probability distribution of $\mathbf{Z}$ after the truncation selection is given by

$$F_{\mathbf{Z}}(\mathbf{z}) = P[\mathbf{Z} \leq \mathbf{z}] = P[G(\mathbf{z}) \geq \theta] \tag{5.26}$$

where $G(\mathbf{z})$ denotes the random variable of total fitness of the individuals with the subsolution $\mathbf{z}$, and θ is a real number (i.e., fitness value) such that

$$\tau = \int_{g(\mathbf{y}) \geq \theta} f_{g(\mathbf{Y})}(g(\mathbf{y}))dg(\mathbf{y}). \tag{5.27}$$

where $g(\mathbf{y})$ is the fitness of the individual $\mathbf{y}$. Since the distribution of $G(\mathbf{z})$ is modeled by $G(\mathbf{z}) \sim \mathcal{N}(\mu_{G(\mathbf{z})}, \sigma_c^2)$ (see Eq. (5.11)), we get

$$F_{\mathbf{Z}}(\mathbf{z}) = P[G(\mathbf{z}) \geq \theta] = \Phi \left(\frac{\mu_{G(\mathbf{z})} - \theta}{\sigma_c} \right) \tag{5.28}$$

where $\Phi(x)$ denotes the cumulative normal distribution function with zero mean and unit variance.

The difference of $(\mu_{G(\mathbf{z})} - \theta)$ almost remains constant regardless of problem size because it can be looked upon as a kind of average fitness contribution of $\mathbf{z}$. In general, the collateral noise σ_c increases with the problem size. It means that the ratio is very small for moderate-to-large problems. There is a linear approximation of normal distribution: $\Phi(x) \approx (1/2) + (x/\sqrt{2\pi})$ for small x. Therefore, Eq. (5.28) can be approximated as follows:

$$F_{\mathbf{Z}}(\mathbf{z}) = \frac{1}{2} + \frac{\mu_{G(\mathbf{z})} - \theta}{\sqrt{2\pi}\sigma_c}. \tag{5.29}$$

The marginal distribution of particular variables Z_1 and Z_2 can be computed by

$$F_{Z_1,Z_2}(z_1, z_2) = F_{Z_1,Z_2,Z_3,\cdots,Z_k}(z_1, z_2, \infty, \cdots, \infty)$$
$$= \frac{1}{2} + \frac{\mu_{G(z_1,z_2)} - \theta}{\sqrt{2\pi}\sigma_c} \tag{5.30}$$

where $\mu_{G(z_1,z_2)}$ is the average fitness of the individuals containing the partition $\mathbf{Z}$ with $Z_1 = z_1$ and $Z_2 = z_2$. Since the numerator $(\mu_{G(z_1,z_2)} - \theta)$ also comes under a fitness contribution (of the sub-block $\{z_1, z_2\}$), the probability distribution is clearly dominated by the denominator $\sqrt{2\pi}\sigma_c$. In other words, the joint probability distribution of Z_1 and Z_2 is inversely proportional to the collateral noise. Thus, the second-order joint moment is proportional to the inverse of collateral noise, viz., $E[Z_1 Z_2] \propto \frac{1}{\sigma_c}$. From the relationship between correlation coefficient and second joint moment given in Eq. (5.25), we get

$$\rho_{Z_1,Z_2} \propto \frac{1}{\sigma_c}. \tag{5.31}$$

By employing Eq. (5.31), Eq. (5.24) can be rewritten as

$$N_s^{crit} = \Theta(\sigma_c^{2.5}). \tag{5.32}$$

Since the collateral noise variance is proportional to the problem size (see Eq. (5.12)), viz., $\sigma_c^2 \propto n$, Eq. (5.32) becomes the following form

$$N_s^{crit} = \Theta(n^{1.25}). \tag{5.33}$$

In other words, the critical population size for discovering dependency between two variables without parents grows quasi-linearly with the size of the problem.

Let us extend the result of Eq. (5.33) to the general case – Z_1 already has multiple parents. If the arcs $X_3 \to X_1$ to $X_k \to X_1$ are already present, the condition for adding the arc $Z_2 \to Z_1$ is clearly given by the following inequality [89, 91]:

$$BIC(Z_1|Z_2, \cdots, Z_k) > BIC(Z_1|Z_3, \cdots, Z_k). \tag{5.34}$$

Naturally, Eq. (5.34) can be rewritten as

$$-N_s h(Z_1|Z_2, \cdots, Z_k) - (k+1)\lambda \ln(N_s) > -N_s h(Z_1|Z_3, \cdots, Z_k) - k\lambda \ln(N_s). \tag{5.35}$$

Setting $I' = h(Z_1|Z_3, \cdots, Z_k) - h(Z_1|Z_2, \cdots, Z_k)$, we have

$$N_s - \frac{\lambda}{I'} \ln(N_s) > 0. \tag{5.36}$$

The critical population size with regard to multiple parents is given by

$$N_s^{crit} - \frac{\lambda}{I'} \ln(N_s^{crit}) = 0. \tag{5.37}$$

From Eqs. (5.19) and (5.37), it follows that the growth of the critical population size N_s^{crit} depends rather strongly on the growth of the mutual information I' [89,91]. We now focus on the growth of I' with the size of the problem.

Let us first consider the case in which Z_1 has only one parent Z_3. In the investigated interaction model (i.e., $Z_2 \to Z_1 \leftarrow Z_3$), the conditional mutual information $I(Z_1; Z_2|Z_3)$ is always less than $I(Z_1; Z_2)$. This is because the process Z_3 plays a role in restricting the range of Z_1 (as in Markov chain models), and thus only partial mutual information of Z_1 and Z_2 is measured. The relationship can be extended to the general case. The conclusion is that the conditional mutual entropy decreases as the number of parents grows. That is, we have $I' \propto I/k$. Thus, the growth of the critical population size in case of multiple parents depends on the product $kn^{1.25}$.

We have derived the result on the BIC employing one component (i.e., $K = 1$). Let us consider the multiple mixture component BIC of Eq. (5.6). In the model selection phase, the structure of the probabilistic model is learned by proportionally combining the BIC results gathered from all the mixture components. It is claimed that the number of individuals involved in each component is inversely proportional to the number of mixture components.

As described in Sect. 5.3.2, employing multiple mixture components provides the ability to partition the search space. Since the mixture model is constructed on the whole population, each component consists of a subset of the population (i.e. subpopulation). The subpopulation sizes are not equal. The size, however, decreases as the mixture components increase. That is, the rate of decrease of the subpopulation size is linearly bounded by the number of components in the mixture. Hence, it supports the claim.

Let there be some unknown dependencies among the mixture components. In order to discover the dependencies, the size of each subpopulation must grow as the product $kn^{1.25}$. Therefore, the critical population size for discovering the problem regularities should grow as

$$N_s^{crit} = \Theta(Kkn^{1.25}) \tag{5.38}$$

where K is the number of mixture components, k is the maximum order of the subproblems, and n is the size of the problem.

Note that three important factors must be considered while computing the population size of EDAs in order to reliably solve problems [89,91]. First, the initial BB supply must be adequate. That is, the population must be large enough to ensure the supply of sufficient raw BBs. Actually, it grows logarithmically with the problem size, i.e., $\Theta(\ln n)$ [7, 41]. Second, the population should guarantee right decisions when selecting the proper solution (from amongst alternatives) to each subproblem so as to ensure the best BB. This requires that the population grows with the square root of the problem size, i.e., $\Theta(\sqrt{n})$ [45, 91]. The above two factors are well known in this regard. Third, the requirement that the population must discover the correct probabilistic model forces a growth expressed by $\Theta(Kkn^{1.25})$. Note that the population size that leads to an accurate model is the dominant factor.

If the number of mixture components (i.e., K) and the order of decomposition (i.e., k) are fixed, the growth of the (rBOA) population is bounded above by $c_1 n^{1.25}$ for some positive constant c_1; and, bounded below by $c_2 n^{1.25}$ for some positive constant c_2, where n is the size of the problem. Hence the conclusion is that

$$N = \Theta(n^{1.25}). \tag{5.39}$$

This population growth enables the rBOA to reach the optimum. This is because the growth model involves all the population-sizing elements needed for global convergence.

5.5.3 Convergence Time Complexity

Another important issue in this regard is the convergence time (i.e., the number of generations until convergence). Our interest is in uniformly scaled subproblems. However, we do not have to be concerned with exponentially scaled subproblems since the number of generations until convergence grows linearly with the size of the problem; that is, $\Theta(n)$ [89, 91]. Recall that the convergence time of BOA (i.e., a complex model with multiple interactions) can be accurately modeled by that of UMDA (i.e., a simple model with no interactions) with regard to various decomposable problems where the order of each (uniformly scaled) subproblem is bounded by a constant [89, 91]. In this context, we may conclude that the convergence time of rBOA can be derived from that of real-coded UMDA[6]. It is necessary to examine this claim prior to proceeding further.

Note that a certain EDA incorporating a fixed factorized distribution in accordance with the variable-interaction structure of the problem (i.e. problem's structure) comes under an FDA. As for rBOA, the population given by Eq. (5.39) ensures that the evolving model converges toward a variable-interaction structure of the problem (i.e., problem's structure). That is, the steady-state dynamics of rBOA is identical to that of a continuous FDA (FDA_c). It may be noted that the FDA_c for separable decomposable problems is mathematically equivalent to the $UMDA_c$ because there are no variable overlaps between subproblems [121].

---

[6] It is also known as continuous UMDA ($UMDA_c$).

With regard to the convergence time, on the other hand, there is no difference between truncation selection and tournament selection. This is explained below. The convergence time (i.e., the number of generations until global convergence) of evolutionary algorithms is inversely proportional to the selection intensity that is defined as the expected increase in the average fitness after selection [89]. However, the selection intensity of truncation selection and tournament selection is a constant throughout the run. It means that they behave identically with regard to convergence time.

González *et al.* [42] investigated the convergence time of $UMDA_c$ equipped with tournament selection. It has been proved that the algorithm is able to reach the optimum in case it starts near about the basin of attraction of the problem, and the speed of convergence decreases with a complexity of $\Theta(1/\sqrt{n})$. If $UMDA_c$ begins far from the optimum, it does not work as expected due to the absence of the signal that leads to the optimum. However, this in itself is of no interest because the investigated rBOA incorporates a population that is sufficient to discover the correct model, and thus can reach the optimum. It indicates that the behavior of rBOA can be approximated by that of $UMDA_c$ starting in the proximity of the basin of attraction.

Therefore, the worst-case convergence time complexity is given by

$$T = \Theta(\sqrt{n}). \tag{5.40}$$

5.5.4 Scalability of rBOA

As described in Sect. 5.5.1, the scalability of rBOA is computed by multiplying the population size required for reliably finding the optimum and the number of generations until convergence.

From Eqs. (5.39) and (5.40), the worst-case complexity in terms of the problem size is given by

$$E = \Theta(n^{1.75}) \tag{5.41}$$

where n is the size of the problem.

As a result, the rBOA finds the optimal solution for decomposable problems, with a sub-quadratic scale-up behavior in respect of the problem size.

5.6 Real-valued Test Problems

This section presents real-valued test problems: (additively) decomposable problems and traditional real-valued optimization problems.

5.6.1 Decomposable Problems

Decomposable problems are created by concatenating basis functions of a certain order. The overall fitness is equal to the sum of all the basis functions. Two types of real-valued decomposable problem are presented.

The first problem is a real-valued deceptive problem (RDP) composed of trap functions. The RDP to be maximized is defined by

$$f_{RDP}(\mathbf{y}) = \sum_{i=1}^{m} f_{trap}(y_{2i-1}, y_{2i}) \tag{5.42}$$

where $y_j \in [0, 1]$, $\forall j$, m are the number of subproblems, and f_{trap} is defined in Eq. (5.43) and plotted in Fig. 5.5(a).

$$f_{trap}(y_j, y_{j+1}) = \begin{cases} \alpha, & \text{if } y_j, y_{j+1} \geq \delta, \\ \frac{\beta}{\delta}\left(\delta - \sqrt{\frac{y_j^2 + y_{j+1}^2}{2}}\right), & \text{otherwise.} \end{cases} \tag{5.43}$$

Here, α and β are the global and the local (i.e, deceptive) optimum, respectively, so that α/β indicates the signal to noise ratio (SNR), and δ is the border of attractors.

Note that the trap function is not only flexible but also quite simple because δ controls the degree of BB supply and the SNR is adjusted by α/β. As an interesting characteristic, it retains 2^m optimal plateaus, out of which there is only one global optimum. The optimum is isolated and there is no attractor around the region, thereby not being amenable to hill-climbing strategies (such as mutation) only. It is clear that recombination is essential to efficiently solve the RDP. In other words, a linkage-friendly recombination operator should be included for preventing disruption of (superior) partial solutions (i.e., BBs).

The second problem is a (real-valued) nonlinear, symmetric problem (RNSP) that is constructed by concatenating nonlinear, symmetric functions. The RNSP to be maximized is

$$f_{RNSP}(\mathbf{y}) = \sum_{i=1}^{m} f_{non\text{-}sym}(y_{2i-1}, y_{2i}) \tag{5.44}$$

where $y_j \in [-5.12, 5.12]$, $\forall j$, and $f_{non\text{-}sym}$ is defined in Eq. (5.45) and illustrated in Fig. 5.5(b).

$$f_{non\text{-}sym}(y_j, y_{j+1}) = \begin{cases} 0.0, & \text{if } 1 - \delta \leq y_j, y_{j+1} \leq 1 + \delta, \\ -100(y_{j+1} - y_j^2)^2 - (1 - y_j)^2, & \text{otherwise.} \end{cases} \tag{5.45}$$

Here, δ adjusts the degree of BB supply, and the nonlinear, symmetric function retains the traits of Rosenbrock function presented in Eq. (5.49).

It is important to note that linkage-friendly recombination which is also capable of capturing nonlinear, symmetric interactions is required for effectively solving the RNSP. It is seen that the RNSP provides a real challenge for real-coded optimization algorithms. Moreover, incorporating the mutation operation further helps find the global optimum as the nonlinear, symmetric function (i.e, basis function) is unimodal so that the hill-climbing strategy at any point eventually leads toward its optimum.

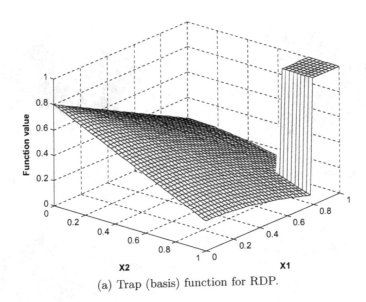

(a) Trap (basis) function for RDP.

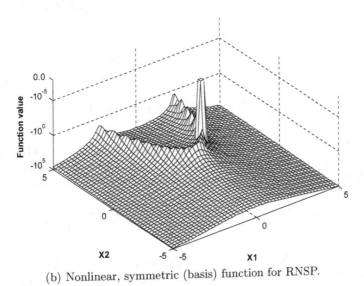

(b) Nonlinear, symmetric (basis) function for RNSP.

Fig. 5.5. Basis functions of decomposable problems.

5.6.2 Traditional Optimization Benchmarks

Four well-known real-valued optimization problems are investigated. Their
two-dimensional versions are illustrated in Fig. 5.6. The task is to minimize

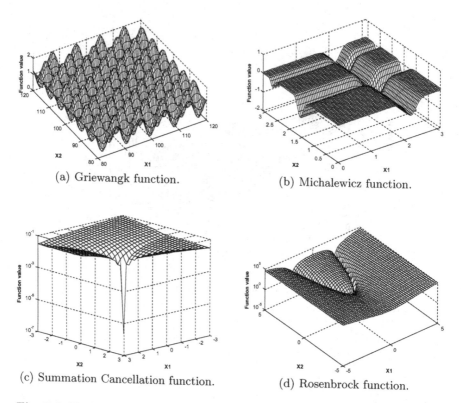

(a) Griewangk function.

(b) Michalewicz function.

(c) Summation Cancellation function.

(d) Rosenbrock function.

Fig. 5.6. Traditional real-valued optimization problems (Two-dimensional forms).

the benchmarks. They have some intriguing characteristics beyond decomposability which most optimization algorithms find hard to negotiate.

The first problem is Griewangk function [113] defined as follows:

$$f_G(\mathbf{y}) = \frac{1}{4000} \sum_{j=1}^{n} (y_j - 100)^2 - \prod_{j=1}^{n} cos\left(\frac{y_j - 100}{\sqrt{j}}\right) + 1 \qquad (5.46)$$

where $y_j \in [-600, 600], \forall j$. It consists of many local optima that prevent optimization algorithms from converging to the global optimum if (fine-grained) gradient information is incorporated. Regardless of its dimensionality, the global optimum is always 0, which is obtained when all (decision) variables are set to 100.

The second problem is Michalewicz function [71] that is given by

$$f_M(\mathbf{y}) = -\sum_{j=1}^{n} sin(y_j) sin^{20}\left(\frac{j \cdot y_j^2}{\pi}\right) \qquad (5.47)$$

where $y_j \in [0, \pi], \forall j$. It also has many suboptimal solutions (albeit to a lesser degree than the Griewangk) and some long valleys along which the minimum value is the same. Thus, gradient information does not lead to better local optima found at the intersections of the channels. The global optimum depends on its dimensionality.

The third problem is Summation-Cancellation function [14] that is formulated as

$$f_{SC}(\mathbf{y}) = \left(10^{-5} + \sum_{j=1}^{n} |y_j + \sum_{i=1}^{j-1} y_i| \right) \Big/ 100 \qquad (5.48)$$

where $y_j \in [-3, 3], \forall j$. The function has (multivariate) linear interactions between variables. Moreover, the optimum is located on a very narrow peak. Thus, it is hard to find the optimal solution without some information on dependencies (of the variables) and dense-searching in the vicinity of the optimum. For any dimension of the problem, the optimum is always 10^{-7}, which occurs when all the variables are 0.

The last problem is Rosenbrock function [96, 99] defined by

$$f_R(\mathbf{y}) = \sum_{j=2}^{n} \left\{ 100 \cdot (y_j - y_{j-1}^2)^2 + (1 - y_{j-1})^2 \right\} \qquad (5.49)$$

where $y_j \in [-5.12, 5.12], \forall j$. It is highly nonlinear and symmetric around quite a flat curved valley. Due to the very small gradient and the strong signal (to solution quality) along the bottom of the valley, it is very hard to find the (global) optimum. Oscillations from one side of the valley to the other is likely unless a starting point is selected in the vicinity of the optimum. The value of the optimum is 0 for any dimensionality. This occurs when the variables are set to 1. No algorithm finds it easy to discover the global optimum of Rosenbrock function.

5.7 Experimental Results and Discussion

This section investigates the ability of rBOA to benefit from the strengths of BOA (i.e., the proper decomposition and the PBBC) in real space. This section also verifies the scalability of rBOA.

5.7.1 Experiment Setup

The performance of rBOA is measured by the average number of (function) evaluations until convergence to the optimum. A comparative study is performed by comparing the solution quality (returned by the fixed number of evaluations) of rBOA with that of EGNA [63], mIDEA [17], and MBOA [81].[7]

---

[7] All the references belong to advanced real-coded, especially direct approach, EDAs.

The references are appropriately tuned in the interest of fair comparison. For instance, the references employ selection and replacement strategies which are identical to those of rBOA.

Among various (un)supervised learning algorithms for accomplishing mixture models, clustering is perceived to be a suitable candidate in terms of computational efficiency [6, 20]. In general, EDAs employ a partitional approach that endeavors to group a set of multi-dimensional data into a number of subsets. Promising examples include K-means algorithm [47] and randomized leader algorithm (RLA) [20, 47]. Their mechanisms are briefly described below.

The K-means algorithm divides data samples into K nonempty subsets. The centroids of the current clusters are computed, and each sample is reassigned to the closest cluster. The process iterates until no more new assignment occurs [47]. In the RLA, each randomly chosen sample belongs to the nearest cluster whose leader is at a distance (to the sample) that is below a given threshold. If there is no such cluster, the sample becomes the leader of a new cluster. The result is obtained after going over the samples exactly once [20, 47].

Note that the RLA is faster than the K-means. The former is somewhat less accurate. Moreover, the frequency with which mixture models are used (in model selection) is lower than that in model fitting. Thus, the K-means algorithm[8] and the RLA (with a threshold value of 0.3) are appropriate candidates for model selection and model fitting, respectively.

Model fitting and model sampling are carried out on the basis of maximal compound subproblems in view of their efficiency for large decomposable problems. Moreover, normal probability distribution has been employed due to its inherent advantages – close approximation and simple analytic properties. Truncation selection that picks the top half of the population and the BIC of Eq. (5.6) whose regularization parameter λ is 0.5 have been invoked for learning a probabilistic model. The renewal policy replaces the worst half of the population with the newly generated offspring (i.e., elitism-preserving replacement). Since no prior information about the problem structure is available in practice, we set $|\mathbf{Y}| - 1$ for the number of allowable parents (i.e., no constraint in the model selection).

Each experiment is terminated when the optimum is found or the number of generations reaches 200. All the results were averaged over 100 runs.

5.7.2 Results for the rBOA Performance

Figure 5.7 shows the average number of evaluations that rBOA performs to find the optimum of RDP with $\alpha = 1.0$, $\beta = 0.8$, $\delta = 0.8$, and n ranging from 10 to 100. The figure also shows results for RNSP with $\delta = 0.2$ and $n = 10$

---

[8] A promising number of clusters (i.e., mixture components K) empirically obtained for each problem is used for model selection.

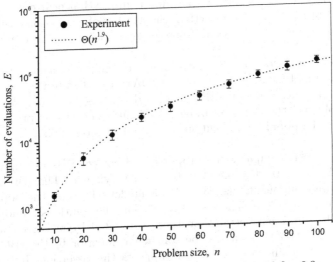

(a) Results for f_{RDP} with $\alpha = 1.0$, $\beta = 0.8$, and $\delta = 0.8$.

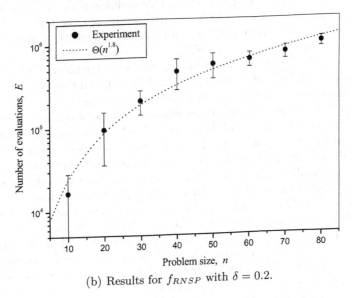

(b) Results for f_{RNSP} with $\delta = 0.2$.

Fig. 5.7. Performance of rBOA on decomposable problems.

to 60. The population size supplied is empirically determined by a bisection method so that the optimum is found [88,89]. In Fig. 5.7, it is seen that the results for the RDP and the RNSP are closely approximated (fitted) by $\Theta(n^{1.9})$ and $\Theta(n^{1.8})$, respectively. Thus, rBOA can solve (additively) decomposable problem of bounded difficulty with a sub-quadratic (but near-quadratic) complexity. In other words, the growth of the number of evaluations with the size of problem (i.e, scalability) seems to be sub-quadratic. A detailed investigation is found in Sect. 5.7.3.

Figure 5.8 provides a comparative study of the performance of rBOA and references (i.e., EGNA, mIDEA, and MBOA) as applied to the decomposable problems (i.e., RDP and RNSP). Since a decomposable problem consists of m subproblems, the effective problem difficulty tends to be proportional to m. Hence, the population is supplied by a linear model, viz., $N = r \cdot m$, for simplicity.

Figure 5.8(a) compares the proportion of correct BBs as applied to the RDP with $\alpha = 1.0$, $\beta = 0.8$, $\delta = 0.8$, and varying m. The rBOA employs one mixture component, viz., $K = 1$, for model selection. The population is supplied by $N = 100m$. The results show that the solutions found by rBOA and MBOA are much better than those computed by mIDEA and EGNA. Although the MBOA seems to be somewhat superior to the rBOA, it has no statistical significance. Table 5.1 supports this assertion. It is also seen that the rBOA and the MBOA achieve stable quality of solutions while the performance of mIDEA and EGNA rapidly deteriorates as the problem size increases. From Figs. 5.7(a) and 5.8(a), it is clear that the scale-up behavior of rBOA and MBOA is sub-quadratic for the RDP; while the mIDEA and the EGNA have an exponential scalability.

Figure 5.8(b) depicts the BB-wise objective function values returned by the algorithms when applied to the RNSP with $\delta = 0.2$ and varying m. Mixture models for model selection use three mixture components ($K = 3$). A linear model, viz., $N = 200m$, is used for supplying the population. As in the RDP, it is seen that the performance of rBOA and MBOA remains uniform irrespective of the problem size. It can mean that they have a sub-quadratic scalability for the RNSP. However, the results show that the rBOA outperforms the MBOA quite substantially with regard to the quality of solution. This consequence is clearly seen in the statistical test in Table 5.1. It is also observed that the mIDEA and the EGNA find solutions of unacceptable quality as the problem size increases and their scalabilities obviously become exponential.

From Figs. 5.7 and 5.8 and Table 5.1, we may conclude that the rBOA finds a better solution with a sub-quadratic scale-up behavior for decomposable problems than does the MBOA, the mIDEA, and the EGNA, especially as the size and difficulty of problems increase.

Table 5.2 compares the solutions found by the algorithms as applied to the well-known real-valued optimization problems depicted in Table 5.1. Three mixture components are employed for all the benchmarks. However, any number of components is acceptable for Griewangk and Michalewicz functions as there is no interaction between variables. The results show that the MBOA is superior to the rBOA, the mIDEA, and the EGNA (they find acceptable solutions, however) for the Griewangk function because it can capture some knowledge about independence as well as overcome numerous traps (i.e., local

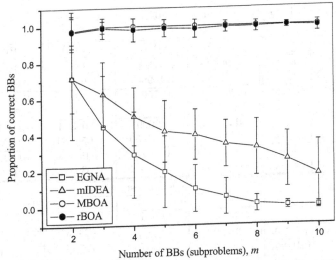

(a) Performance of algorithms on f_{RDP} with $\alpha = 1.0$, $\beta = 0.8$, $\delta = 0.8$ and various m.

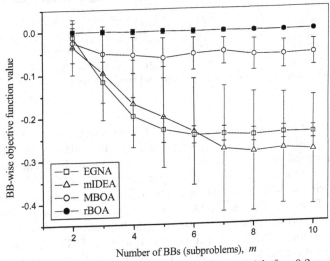

(b) Performance of the algorithms on f_{RNSP} with $\delta = 0.2$ and different m.

Fig. 5.8. Comparison results of algorithms on decomposable problems.

optima) due to the kernel distributions. In the Michalewicz function, the performances of MBOA and rBOA are comparable, and both algorithms outperform the EGNA and the mIDEA. It means that the EGNA and the mIDEA

Table 5.1. Performance comparison of algorithms on f_{RDP} and f_{RNSP}.

Problem	Measure	EGNA	mIDEA	MBOA	rBOA
RDP	μ_{QoS}	0.196000	0.418000	1.0	0.988000
($m = 5$)	σ_{QoS}	0.197949	0.169900	0.0	0.047497
RDP	μ_{QoS}	0.002000	0.175000	1.0	0.992000
($m = 10$)	σ_{QoS}	0.019900	0.187283	0.0	0.030590
RNSP	μ_{QoS}	−0.229916	−0.200973	−0.063843	−0.001384
($m = 5$)	σ_{QoS}	0.030276	0.136850	0.056469	0.005965
RNSP	μ_{QoS}	−0.238623	−0.299768	−0.056143	−0.001456
($m = 10$)	σ_{QoS}	0.017609	0.111364	0.030395	0.002651

Statistical t-test					
Test case		RDP		RNSP	
		$m = 5$	$m = 10$	$m = 5$	$m = 10$
rBOA − EGNA		38.30†	273.20†	71.72†	110.78†
rBOA − mIDEA		32.80†	41.92†	14.45†	13.59†
rBOA − MBOA		−2.51	−1.99	11.10†	13.34†
MBOA − EGNA		40.41†	499.00†	27.18†	33.51†
MBOA − mIDEA		34.08†	43.83†	14.71†	10.43†
mIDEA − EGNA		8.10†	9.05†	2.19	−1.97

Problem	Statistical order
RDP ($m = 5, 10$)	rBOA $\sim$ MBOA $\succ$ mIDEA $\succ$ EGNA
RNSP ($m = 5, 10$)	rBOA $\succ$ MBOA $\succ$ mIDEA $\sim$ EGNA

† The value of t is *significant* at $\alpha = 0.01$ by a paired, two-tailed test. The symbols $\succ$ and $\sim$ represent *dominance* and *indifference* between algorithms.

fail to discover independent interactions between variables. It is also seen that the EGNA and the rBOA are quite superior to the mIDEA and the MBOA in the Cancellation function. Although all the algorithms can successfully capture the information about linear interactions, the EGNA achieves the best performance due to its inherent efficiency when it comes to single-peak functions. Even though the rBOA traverses multiple regions of the unimodal function, its performance is acceptably high. It is important to note that the rBOA outperforms the MBOA, the mIDEA, and the EGNA in the case of the Rosenbrock function whose optimum is hard to find. Further, the performance of MBOA and EGNA is very poor. This is explained below.

The variables of the Rosenbrock function strongly interact around a curved valley. Also, the function is symmetric. It is clear that incorrect factorizations (i.e., no dependencies between variables) are encountered at an early stage of the algorithms. Due to the incorrect structure, they try to solve the problems by treating the variables in isolation. Of course, finding an optimum in this way is difficult because any given algorithm does not cross the intrinsic barrier. After a few generations, however, individuals start to collect around

Table 5.2. Performance of algorithms on real-valued benchmarks ($n = 5$).

Problem	Measure	EGNA	mIDEA	MBOA	rBOA
Griewangk	μ_{QoS}	0.061968	0.067873	0.003258	0.065993
($N = 2000$)	σ_{QoS}	0.016287	0.018634	0.005205	0.017604
Michalewicz	μ_{QoS}	−4.637647	−4.613430	−4.687653	−4.687640
($N = 500$)	σ_{QoS}	0.013388	0.076301	0.005857	0.000044
Cancellation	μ_{QoS}	0.000034	0.014854	0.001654	0.000557
($N = 100$)	σ_{QoS}	0.000122	0.006420	0.001663	0.000740
Rosenbrock	μ_{QoS}	2.141721	0.003518	0.664121	0.000177
($N = 3000$)	σ_{QoS}	0.182596	0.017894	0.521631	0.001283

Statistical t-test				
Test case	Griewangk	Michalewicz	Cancellation	Rosenbrock
EGNA − rBOA	−1.70	37.17[†]	−6.69[†]	116.64[†]
mIDEA − rBOA	0.74	9.68[†]	21.97[†]	1.83
MBOA − rBOA	−33.77[†]	0.00	5.72[†]	12.67[†]
EGNA − MBOA	32.76[†]	37.16[†]	−9.58[†]	25.49[†]
mIDEA − MBOA	33.07[†]	9.68[†]	19.83[†]	−12.65[†]
EGNA − mIDEA	−2.35	−3.30[†]	−22.91[†]	115.46[†]

Problem	Statistical order
Griewangk	MBOA ≻ rBOA ∼ mIDEA ∼ EGNA
Michalewicz	rBOA ∼ MBOA ≻ EGNA ≻ mIDEA
Cancellation	EGNA ≻ rBOA ≻ MBOA ≻ mIDEA
Rosenbrock	rBOA ∼ mIDEA ≻ MBOA ≻ EGNA

[†] The value of t is *significant* at $\alpha = 0.01$ by a paired, two-tailed test. The symbols ≻ and ∼ represent *dominance* and *indifference* between algorithms.

the curved valley. At this time, the rBOA can easily capture such a nonlinear, symmetric dependency due to mixture models. On the other hand, the mIDEA can cope with the cancellation effect (arising from symmetry) to some extent by clustering in the overall problem space. However, the MBOA does not deal successfully with the situation because finding a promising set of split boundaries so as to cross the barrier is very difficult. In addition, the EGNA finds it impossible to overcome the hurdles by a (simple) single-peak model.

From Table 5.2, it can be concluded that the rBOA finds good solutions to complicated problems in terms of dependencies (of decision variables) while achieving comparable or acceptable solutions to others.

As a result, the rBOA achieves the optimal solution with a sub-quadratic scale-up behavior for decomposable problems. Note that the sub-quadratic scalability is solely due to proper decomposition brought about by correct factorization and the PBBC realized by the subspace-based model fitting and model sampling.

Moreover, the rBOA finds better solutions for decomposable problems and acceptable (or even better) solutions to traditional real-valued optimization benchmarks, than those found by the state-of-the-art real-coded EDAs.

5.7.3 Verification of rBOA Scalability

Figures 5.9 and 5.10 show the experimental results for decomposable problems (i.e., RDP and RNSP). The population size N supplied is empirically sought by a bisection method [89] such that the rBOA finds the optimal solution by the estimated (optimal) population size.

Figure 5.9 depicts the average population size and the number of evaluations until convergence to the optimum when the rBOA with $K = 1$ is applied to the RDP with $\alpha = 1.0$, $\beta = 0.8$, and $\delta = 0.8$. The size of the tested problem varies from $n = 10$ to 100. Figure 5.10 also illustrates the same performance measures as the rBOA with $K = 3$ is applied to the RNSP with $\delta = 0.2$. The size n of the problem varies from 10 to 60.

The results in Figs. 5.9(a) and 5.10(a) show that the population-sizing model $\Theta(n^{1.25})$ (i.e., population-complexity) (required for correctly discovering the problem regularities) is in close agreement with the experimental results. In Figures 5.9(b) and 5.10(b), it is also seen that the theoretical model of rBOA scalability, viz., $\Theta(n^{1.75})$, is consistent with experimental results. More accurate approximations for the RDP and the RNSP are obtained by $\Theta(n^{1.9})$ and $\Theta(n^{1.8})$ respectively (see Fig. 5.7). The discrepancies arise from the effects of finite population on the convergence time and the assumptions made for tractable analysis. The close agreement with actual performance is a noteworthy feature in this regard.

We have often insisted that the analysis is also applicable to exponentially scaled problems where the assumptions made for uniformly scaled problems are relaxed. As described in Sect. 5.5.3, the only difference is in the convergence time: the number of evaluations increases linearly with the problem size. It implies that the population dependence remains the same, i.e., $\Theta(n^{1.25})$, and hence the convergence time complexity is $\Theta(n)$. Thus, the scalability of rBOA for exponentially scaled problems is reflected by $\Theta(n^{2.25})$.

Additional experimental studies are reported in this section. On the basis of RDP, an exponentially scaled problem is produced by multiplying subproblems by exponential-scaling constants. The problem is formally stated by

$$\sum_{i=1}^{m} c^i \cdot f_{trap}(y_{2i-1}, y_{2i}) \tag{5.50}$$

where c is a scaling factor.[9]

Fig. 5.11 illustrates the average population size and the number of evaluations as applied to the exponentially scaled problem. It is also seen that all

---

[9] In this experiment, the scaling factor c is set to 2.5.

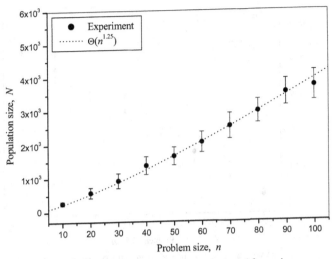

(a) Average population size versus problem size.

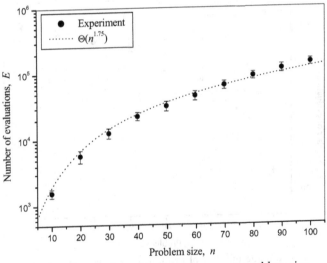

(b) Average number of evaluations versus problem size.

Fig. 5.9. Scalability of rBOA on f_{RDP} with $\alpha = 1.0$, $\beta = \delta = 0.8$, and $n = 10 - 100$.

the experimental results are in close agreement with the theoretical bounds. In other words, it supports the validity of the method of analysis in respect of exponentially scaled problems.

On the basis of the results, we may conclude that the rBOA achieves an optimum with a sub-quadratic scale-up behavior for (uniformly scaled)

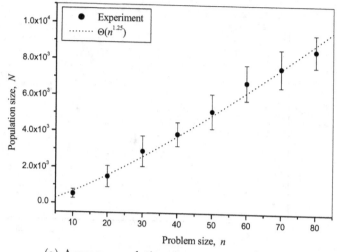

(a) Average population size versus problem size.

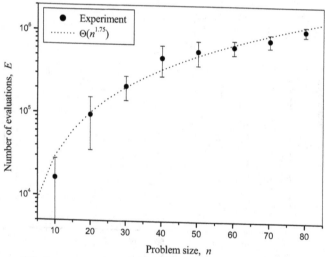

(b) Average number of evaluations versus problem size.

Fig. 5.10. Scalability of rBOA on f_{RNSP} with $\delta = 0.2$ and $n = 10 - 60$.

decomposable problems, and its population size supplied for learning a correct model grows as quasi-linear.

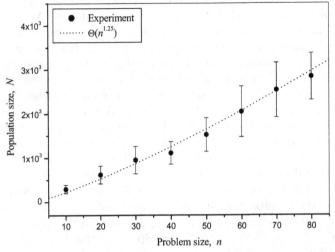

(a) Average population size versus problem size.

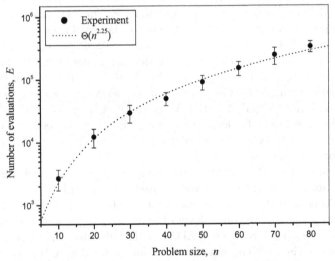

(b) Average number of evaluations versus problem size.

Fig. 5.11. Scalability of rBOA on the exponentially scaled problem.

5.8 Summary

In this chapter, we have presented a real-coded BOA in the form of (advanced) real-coded EDAs. Decomposable problems were the prime targets and sub-quadratic scale-up behavior (of rBOA) was a major objective. This was achieved by proper decomposition (i.e., linkage learning) and probabilistic building-block crossover (PBBC) on real-valued variables. As a step in this direction, Bayesian factorization was performed by means of mixture models, the substructures were extracted from the resulting Bayesian factorization graph (i.e., problem decomposition), and each substructure was fitted by mixture distributions whose parameters were extracted (by estimation) from the subspaces (i.e., subproblems). In the model sampling phase, the offspring was generated by an independent subproblem-wise sampling procedure.

Moreover, the scalability of rBOA has been analyzed for uniformly scaled decomposable problems. The approach has also been valid for exponentially scaled problems. According to the number of mixture components (K), the (maximum) order of decomposition (k), and the size of the problem (n), it has been shown that the population size required for building a correct model (i.e., discovering the problem regularities) increases as $Kkn^{1.25}$ and the number of evaluations until reaching the optimum grows as $Kkn^{1.75}$. That is, the worst-case complexity of rBOA scalability is $\Theta(n^{1.75})$. Thus, the rBOA finds the optimal solution with a sub-quadratic scalability with regard to the problem size.

Experimental studies demonstrated that that the rBOA finds the optimal solution with a sub-quadratic scale-up behavior. The comparative studies exhibited that the rBOA outperforms the up-to-date real-coded EDAs (EGNA, mIDEA, and MBOA) when faced with decomposable problems regardless of inherent problem characteristics such as deception, nonlinearity, and symmetry. Moreover, the solutions computed by rBOA are acceptable in the case of traditional real-valued optimization problems while they are generally better than those found by EGNA, mIDEA, and MBOA. Further, the quality of solutions improves with the degree of problem difficulty. Moreover, the analytic models of rBOA vis-à-vis the population-sizing and the scalability have been verified by the experimental studies.

It is noted that the rBOA learns complex dependencies of variables by means of mixture distributions and estimates the distribution of population by exploiting mixture models at the level of substructures. This allows us to keep options open at the right level of attention throughout the run. In the past, most (advanced) real-coded EDAs used single normal models or mixtures at the level of the problem, but these are unable to capture the critical detail.

More work on the proper number of mixture components and the development of faster mixture models needs to be done. However, rBOA's strategy of decomposing problems, modeling the resulting building blocks, and then searching for better solutions appears to have certain advantages over existing advanced probabilistic model building methods that have been suggested and used elsewhere. Certainly, there can be many alternatives with regard to exploring the method of decomposition, the types of probabilistic models utilized, as well as their computational efficiency, but this avenue appears to lead to a class of practical procedures that should find widespread use in many engineering and scientific applications.

6

Multiobjective Real-coded Bayesian Optimization Algorithm

This chapter presents a competent *multiobjective estimation of distribution algorithm* (MEDA). It solves real-valued multiobjective optimization problems (MOPs) of bounded difficulty quickly, accurately, and reliably: it is a pilot study in this regard. This is the *multiobjective real-coded Bayesian optimization algorithm* (MrBOA). The goal is to fit the real-coded Bayesian optimization algorithm (rBOA) into the multiobjective optimization framework without in any way diluting its unique features. Thus, it can automatically discover problem regularities and then effectively exploit this knowledge for traversing the multiobjective search space in the context of decomposition principle. A new selection scheme has also been developed for improving proximity and promoting diversity of the solutions. It assigns fitness to nondominated individuals on the basis of pure *Pareto ranks* and *crowding distance*; and to others by their ranks and *sharing intensity*. The effect of the Pareto rank is to push all the individuals toward nondominated solutions. It may significantly weaken population diversity, but the problem is overcome by incorporating *crowding distance* and *sharing intensity*. The crowding distance (that has already been investigated elsewhere) is enhanced by considering the distance of a solution from its successor along each of the objectives. It can promote diversity of nondominated individuals due to its ability to discriminate between two equally preferable but crowded individuals and then select the most promising one. The sharing intensity is inversely proportional to the *domination count*, which is the number of individuals that dominates the solutions. It can also preserve population diversity. In other words, it penalizes the dominated individuals by their domination count so that representative individuals are selected. By combining the selection scheme with the rBOA, the MrBOA achieves a Pareto front with improved proximity and diversity.

The chapter is organized as follows. Section 6.1 introduces the central concepts of multiobjective optimization. Section 6.2 briefly reviews genetic and evolutionary algorithms (GEAs) for multiobjective optimization. Section 6.3 outlines MrBOA. The proposed selection scheme is presented in Sect. 6.4.

Chang Wook Ahn: *Advances in Evolutionary Algorithms: Theory, Design and Practice*, Studies in Computational Intelligence (SCI) **18**, 125–151 (2006)
www.springerlink.com   © Springer-Verlag Berlin Heidelberg 2006

Real-valued test MOPs are cited in Sect. 6.5 and experimental results are exhibited in Sect. 6.6. The chapter concludes with a summary in Sect. 6.7.

6.1 Multiobjective Optimization

Many real-world problems naturally fall within the purview of *multiobjective* (or *multicriteria*) *optimization problems* (MOPs) consisting of several incommensurable and often conflicting objectives [20, 24, 36, 68, 122, 123]. Without loss of generality, a general MOP can be formulated as follows:

$$\begin{aligned} \text{minimize} \quad & \mathbf{z} = \mathbf{f}(\mathbf{y}) = (f_1(\mathbf{y}), f_2(\mathbf{y}), \cdots, f_M(\mathbf{y})) \\ \text{subject to} \quad & \mathbf{e}(\mathbf{y}) = (e_1(\mathbf{y}), e_2(\mathbf{y}), \cdots, e_J(\mathbf{y})) \geq 0 \end{aligned} \quad (6.1)$$

where $\mathbf{y} = (y_1, y_2, \cdots, y_n) \in \Omega$, $\mathbf{z} = (z_1, z_2, \cdots, z_M) \in \Lambda$. Here, $\mathbf{y}$ is the decision vector, Ω denotes the decision space, $\mathbf{z}$ is the objective vector, and Λ denotes the objective space. The set of decision vectors $\mathbf{y}$ that satisfy the constraints $\mathbf{e}(\mathbf{y}) \geq 0$ is defined to be the feasible set Ω_f, and the image of Ω_f in the objective space is defined to be the feasible region Λ_f.

Due to the interdependence of the objectives, MOPs normally have a set of alternative solutions. These solutions are optimal in the sense that no solution is superior to them in an overall sense because no objective can be improved without losing on the others. The set of solutions is known as *Pareto-optimal set* or *nondominated set*. With regard to a set $\mathcal{A} \subseteq \Omega_f$, the Pareto-optimal set (or nondominated set) that consists of alternative solutions such that no objective can be improved without, at the same time, degrading the others is mathematically defined by

$$\mathcal{Q} = \{\mathbf{y}^0 \in \mathcal{A} | \nexists \, \mathbf{y}^1 \in \mathcal{A} : \mathbf{y}^1 \succ \mathbf{y}^0\} \quad (6.2)$$

where $\mathbf{y}^1 \succ \mathbf{y}^0$ indicates that the solution $\mathbf{y}^1$ (*Pareto*) *dominates* the solution $\mathbf{y}^0$. That is,

$$\forall i : f_i(\mathbf{y}^0) \geq f_i(\mathbf{y}^1) \wedge \exists j : f_j(\mathbf{y}^0) > f_j(\mathbf{y}^1). \quad (6.3)$$

The image of the Pareto-optimal set under the feasible objective space is defined to be the *Pareto (optimal) front*. It is defined by

$$\mathcal{F} = \{(f_1(\mathbf{y}^0), f_2(\mathbf{y}^0), \cdots, f_M(\mathbf{y}^0)) | \mathbf{y}^0 \in \mathcal{Q}\}. \quad (6.4)$$

Note that the goal of multiobjective optimization is to find the *global Pareto-optimal set* $\mathcal{Q}^*$; in other words, to locate the *global* (or *true*) *Pareto front* $\mathcal{F}^*$ of the nondominated solutions. They are defined by the Pareto-optimal set and the Pareto front on the entire feasible set Ω_f and region Λ_f, respectively.

However, achieving the goal is not practical since there can be an infinite number of Pareto-optimal solutions. Therefore, the down-to-earth goal is to find representative nondominated solutions of the true Pareto front while maintaining a good spread of solutions over the front [18–20, 24, 29, 38, 60].

6.2 Multiobjective Genetic and Evolutionary Algorithms

Multiobjective genetic and evolutionary algorithms (MGEAs) have attracted due attention of late due to their ability to search for multiple solutions in parallel (so that a family of feasible solutions to the problem is found) as well as handle complex features such as discontinuity, multimodality, and disjoint objective spaces [29, 68, 123]. The growing interest in difficult, higher dimensional problems has spurred the growth of MGEAs for over a decade. In general, MGEAs can be divided into two categories – *population-based* and *probability distribution-based* (or *EDA-based*) approaches.

The population-based approach strives to deal with MOPs by allowing population-based GEAs the ability to handle multiple objectives. It can be further classified into two categories: *simple* and *advanced*.

The simple approach attempts to improve proximity of the Pareto front by exploiting the domination information of individuals and maintain diversity of the population by employing a sharing strategy. However, the approach often looses good solutions found so far due to lack of elitism and fails to maintain diversity in the population due to the difficulty in properly specifying the sharing parameter. Multiobjective genetic algorithm (MOGA) [37], niched Pareto genetic algorithm (NPGA) [54], and nondominated sorting genetic algorithm (NSGA) [109] are included in this category.

The MOGA assigns the rank of a particular individual by adding one to the number of individuals by which it is dominated. All the nondominated individuals are preferentially managed by assigning the best rank; while dominated solutions are penalized by the population density of the corresponding region of the tradeoff surface. The fitness assignment is performed by usually linearly interpolating the ranks from the nondominated (i.e, the best) to the most dominated (i.e., the worst) individuals. The MOGA also uses a niche-formation method to distribute the population over the Pareto front, which maintains the diversity in the population.

The NPGA harmonizes tournament selection with the concept of Pareto dominance. Instead of allowing direct competition between two individuals, a comparison set consisting of other individuals in the population intervenes in the tournament. If one of the two competitors is dominated by any member of the comparison set and the other is not, then the latter is chosen as the winner of the tournament. When both competitors are either dominated or nondominated, the tournament result is determined by fitness sharing: the individual of the sparsest population density in its niche is selected. Although quite a large population size is inevitable, the selection noise can be tolerated by emerging niches in the population.

The NSGA classifies individuals into several categories based on the concept of nondomination. To provide an equal signal for reproduction, all nondominated individuals are classified into the first (i.e., nondominated) front and assigned the same dummy fitness value. This group of individuals is then ignored and the next front is extracted in the same manner. The process is repeated until all individuals in the population are classified. This is known as the *nondominated sorting*. To maintain the diversity of the population, these solutions are shared with their dummy fitness values by a niching method.

To compensate the inherent defects of the simple approach, an advanced approach has been devised. It endeavors to harmonize with elitism and to take into account domination and density information at the same time, thereby improving both proximity and diversity of nondominated solutions. However, the approach may not be efficient for some complicated problems since it does not pay much attention to linkage-friendly recombination. It is a key component (of GEAs) for growing and assembling good partial solutions (i.e., BBs) toward global optima. Strength Pareto evolutionary algorithm (SPEA) [122, 123], SPEA-II [124], NSGA-II [29], and rank-density-based genetic algorithm (RDGA) [68] are representative schemes.

The SPEA archives nondominated solutions found thus far. In the external set (called *archive*), the fitness of an individual is determined by a proportional number of individuals which are covered by the individual. The proportional number is defined as *strength*. For an individual of the non-external set (i.e., the population), its fitness is calculated by adding "1" to the total sum of the strengths of all the external members that cover the individual. This mechanism simultaneously performs two tasks on the individuals preferably closer to the Pareto front; and the population diversity is maintained without any explicit sharing. The SPEA also incorporates a clustering procedure in order to keep the size of the external set without affecting its unique characteristics.

The SPEA-II is an enhanced version of SPEA. In the SPEA-II, each individual in both the (elitist) archive and the population (i.e., non-external set) is assigned a strength value by the number of individuals it dominates. The rank (i.e., raw fitness) of an individual is computed by summing the strengths of the individuals that dominate the individual. To discriminate between individuals which have equal ranks, the density information is estimated by the kth nearest neighbor method. The (final) fitness is defined as the sum of rank and density values. In addition, a truncation method is used in the archive in order to maintain the number of elitists.

The NSGA-II combines a fast nondominated sorting approach and a simple crowding distance assignment method in the earlier NSGA framework. Also, elitism is involved in the selection phase. The nondominated sorting scheme can rapidly discover the *Pareto ranks* (i.e., *domination ranks*) of individuals. As a diversity-preserving strategy, a dynamic crowding scheme is applied to effectively carry out density estimation. The NSGA-II prefers the individuals that are close to the true Pareto front and exist in some coarser region (i.e., low density) when they belong to the same front. Moreover, the idea of elitism is all about equally preferable nondominated solutions.

The RDGA exploits the ranking scheme of MOGA in the similar context of the strength of SPEA and SPEA-II. The final rank of an individual is defined as the sum of the rank values of all the individuals that dominate it. In addition, a modified adaptive cell density evaluation scheme is employed for discriminating between the identically ranked individuals. It transforms any MOP into a biobjective optimization problem over the rank-density domain. After the conversion, fitness assignment is efficiently fulfilled by a simple MGEA such as a vector evaluated genetic algorithm (VEGA) [103].

The EDA-based approach concentrates on effectively combining the strengths of the state-of-the-art MGEAs with the EDAs' ability of automatic discovery and exploitation of problem regularities. The approach generally outperforms the population-based approach with regard to both proximity and diversity, by virtue of its efficient fitness assignment policy of competent MGEAs and linkage-friendly recombination of EDAs. In this sense, the EDA-based approach is coming into limelight of late. Bayesian mutiobjective optimization algorithm (BMOA) [82], multiobjective (hierarchical) Bayesian optimization algorithm (m(h)BOA) [59,60], and multiobjective mixture-based iterative density-estimation evolutionary algorithm (MIDEA) [18,19] are some leading techniques.

The BMOA merges a ϵ-dominance selection into mixed Bayesian optimization algorithm (MBOA) for multiobjective optimization. In the selection, no two neighboring individuals within an ϵ-distance are nondominated, and the survival of dominated solutions depends on the number of individuals by which they are dominated. Thus, the selection operator enables nondominated solutions to ensure proximity (i.e., convergence to the true Pareto front) and diversity (i.e., a good approximation of the front).

The m(h)BOA combines the simple and robust selection scheme of NSGA-II with the BB identification and mixing capabilities of (h)BOA for multiobjective optimization. Due to the synergistical integration, it can effectively solve MOPs of (hierarchically) bounded difficulty.

The MIDEA transplants a *diversity-preserving selection* into an iterative density-estimation evolutionary algorithm (IDEA) with mixture probability distributions in the multiobjective optimization framework. The selection operator can flexibly strike a balance between proximity and diversity in the resulting approximation set through a single control parameter. The use of mixture distributions obtained by means of clustering the objective space further stimulates the diversity of solutions. In addition, the diversity-preserving selection reconciled with the elitism can prevent diversity degeneration.

6.3 Multiobjective Real-coded Bayesian Optimization Algorithm

This section describes the multiobjective real-coded Bayesian optimization algorithm (MrBOA) as an effective tool for solving MOPs. The aim is to extend the single-objective real-coded Bayesian optimization algorithm (rBOA) into the realm of multiobjective optimization.

Without loss of generality, it is assumed that all the objectives in the MOP are to be minimized. Further, real-valued multiobjective optimization problems (RMOPs) are the main targets since the (single-objective) rBOA treats the continuous domain quite satisfactorily.[1] In this regard, the rBOA is extended into the realm of multiobjective optimization without in any way diluting its unique features. Generously drawing on the procedures of rBOA (see Section 5.2), the following pseudo-code provides an outline of the MrBOA:

STEP 1. INITIALIZATION
    Randomly generate initial population $\mathcal{P}$
STEP 2. SELECTION
    Select a set of promising candidates $\mathcal{S}$ from $\mathcal{P}$
    2.1. RANKING
        Domination ranks $\mathcal{R}$ are found by the nondominated sorting
    2.2. ADAPTIVE SHARING
        Sharing intensity $\mathcal{I}$ are computed by an adaptive sharing
    2.3. DYNAMIC CROWDING
        Crowding distances $\mathcal{D}$ are computed by a dynamic crowding
    2.4. FITNESS ASSIGNMENT
        Fitness of individuals are assigned by $\mathcal{R}$, $\mathcal{I}$, and $\mathcal{D}$
    2.5. ELITISM
        The elitist solutions are selected
STEP 3. LEARNING
    Learn a probabilistic model $\mathcal{M}$ from $\mathcal{S}$ using a metric (and constraints)
STEP 4. SAMPLING
    Generate a set of offspring $\mathcal{O}$ from the estimated probability distribution
STEP 5. REPLACEMENT
    Create a new population $\mathcal{P}$ by replacing some individuals of $\mathcal{P}$ with $\mathcal{O}$
STEP 6. TERMINATION
    If the termination criteria are not satisfied, go to STEP 2

All the procedures except for the selection (i.e., STEP 2) are the same as those of rBOA. Hence, the selection procedure imparts to rBOA the capability to handle multiple objectives, and the (probabilistic) model learning (i.e., building) and sampling technique of rBOA provide MrBOA with necessary tools for discovering problem regularities and achieving the maximum BB-wise mixing rate in multiobjective optimization.

In general, the MGEAs are characterized by the selection strategy. This is because the purpose of selection in the MGEAs is to choose individuals that can lead individuals to the true Pareto front $\mathcal{F}^*$ while maintaining a good spread of the solutions. The selection method of STEP 2 is described in the next section.

---

[1] However, different types of problems can be taken into consideration because the rBOA can be replaced by any competent GEA while directly incorporating the proposed selection method in Section 6.4.

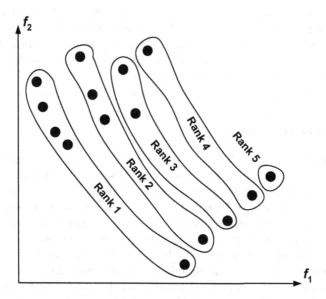

Fig. 6.1. Example of domination ranks.

6.4 Selection Strategy

This section presents the proposed selection scheme that promotes high proximity as well as diversity-preservation. It consists of five components described below.

6.4.1 Ranking

In MGEAs, ranking is a fundamental element since it is closely related to fitness assignment to individuals. Many ranking schemes have been developed for achieving close convergence and uniform spread to the true Pareto front $\mathcal{F}^*$ [29, 37, 68, 82, 123, 124].

We employ nondominated sorting method (i.e., pure Pareto ranking scheme) of NSGA-II, due to its simplicity and effectiveness. It decides domination ranks $\mathcal{R}$ of individuals in such a manner that all the nondominated individuals in the population are assigned "rank 1" (known as the *first* (Pareto) front[2] $\mathcal{F}_1$) and removed from temporary assertion, then a new set of nondominated individuals is assigned "rank 2" (viz., the *second* front $\mathcal{F}_2$), and so forth. A solution (i.e., individual) with a lower rank is always preferred. An example of the Pareto ranking is illustrated in Fig. 6.1.

In the NSGA-II, a simple but efficient crowding method for diversity preservation is applied to the individuals on the basis of identical fronts.

---

[2] It is also denoted as the *nondominated* front.

It computes crowding distance (as a density estimate) in order to discriminate among solutions with equal domination ranks. However, its effect on the selection of individuals is definitely secondary to that of their domination ranks. Thus, some information that can facilitate diversity preservation must be more actively brought in for selecting promising individuals. In this regard, *adaptive sharing* and *dynamic crowding* described below are very helpful.

6.4.2 Adaptive Sharing

The adaptive sharing method effectively estimates density information of individuals. It is based on the *domination count* [18,36]. The domination count of an individual is defined by the number of individuals in the current population by which it is dominated. The goal of the sharing scheme is to boost the solutions that are less dominated since they generally stand for their dominated solutions. In a similar way, an individual can also incorporate the information about the number of individuals which are dominated by it; that is, a solution that dominates more individuals is preferred [24]. However, it risks being a primary factor. The reason is that a solution dominated by a smaller number of individuals (i.e., smaller domination count) is essentially representative; but a solution that dominates a smaller number of individuals is not necessarily unrepresentative. The latter case commonly occurs whenever the solutions exist around the boundary of each objective. In that region, some individuals may dominate only a few solutions even though their ranks are low. Moreover, this situation frequently occurs as population converges toward the true Pareto front. Consequently, the domination count is sufficient to distinguish representative individuals of a population.

In the existing sharing methods, the performance is strongly governed by the parameter setting. Thus, it cannot achieve a good performance without proper setting. However, the adaptive approach does not require any user-specified parameter. That is, good density information can be adaptively computed by employing the domination count alone. As a measure of density, *sharing intensity* is defined as follows:

$$\mathcal{I}(i) = 1 - \frac{1}{1 + N_{dom}(i)} \tag{6.5}$$

where $\mathcal{I}(i)$ is the sharing intensity of individual i, $N_{dom}(i)$ is the domination count of the individual i.

It assigns zero value to the nondominated individuals and lower values are assigned to the individuals which are less dominated (by other solutions). Hence, an individual with a lower value is always preferred because it serves as a representative of its objective space covered by the solution regardless of the number of dominated individuals in that region. During fitness assignment (in Sect. 6.4.4), it can play an important role in preserving diversity of population by imposing higher penalty to the individuals which are dominated by more

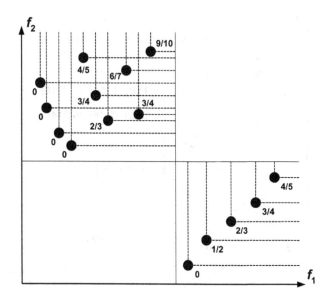

Fig. 6.2. Example of sharing intensity.

individuals because they can be regarded as being crowded (i.e., redundant) in the sense of Pareto optimality. An example of sharing intensity of individuals is given in Fig. 6.2.

In the figure, domination ranks of individuals are the same as those in Fig. 6.1. It is seen that lower values are assigned to individuals closer to non-dominated solutions having smaller number of neighbors (i.e., less crowded). In other words, the individuals that can promote diversity preservation (and improve proximity as a bonus) are preferred.

6.4.3 Dynamic Crowding

Sine the sharing method does not have any effect on nondominated individuals (see Figure 6.2), the dynamic crowding method is applied for performing a good approximation of the nondominated (i.e., first) front $\mathcal{F}_1$. It has been noted that the crowding method of NSGA-II is quite effective in stimulating a diverse representation of the nondominated solutions in $\mathcal{F}_1$. Hence, we employ the crowding method; it has been further enhanced, however.

After sorting individuals according to each objective function value, the crowding method is applied to the first front $\mathcal{F}_1$ in order to discriminate between the nondominated solutions[3]. As a density measure, *crowding distance* is defined by

---

[3] They have the same domination rank "1" and sharing intensity "0".

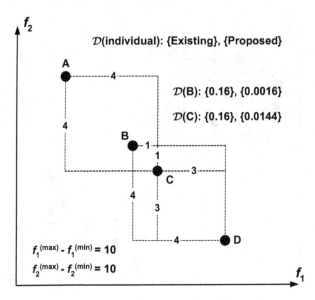

Fig. 6.3. Example of crowding distance.

$$D(\omega_i) = \sum_{k=1}^{v} \frac{\{f_k(\omega_{i+1}) - f_k(\omega_{i-1})\}\{f_k(\omega_i) - f_k(\omega_{i-1})\}}{\left(f_k^{max} - f_k^{min}\right)^2} \qquad (6.6)$$

where ω_i is the ith individual in the sorted set of nondominated solutions, $D(i)$ is the crowding distance of the individual i, $f_k(i)$ is the kth objective function value of the individual i, and f_k^{min} (f_k^{max}) is the minimum (maximum) value of the kth objective function. For each objective, as a matter of course, the first and last individuals are assigned an infinite distance to give an absolute preference to boundary solutions.

An individual with a larger value is always preferred because it is regarded as a less crowded (i.e, representative) individual that can well approximate the nondominated solutions. Thus, the crowding distance can be incorporated with fitness assignment as a sort of penalty function.

On the other hand, its benefit over the crowding of NSGA-II is illustrated in Fig. 6.3. It is seen that the crowding of NSGA-II assigns to individuals B and C the same values so that both individuals are equally preferred. But, it is sufficient if one of them survives for a good approximation of the solutions in $\mathcal{F}_1$ because the two solutions are quite close (i.e., crowded). The proposed method assigns to the individual C a crowding distance that is larger than that assigned to the individual B. That is, C is preferred to B. Therefore, the crowding method provides an efficient framework for further promoting diversity preservation of the nondominated solutions.

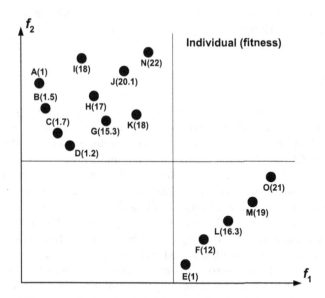

Fig. 6.4. Example of fitness assignment with $\gamma = 20$.

6.4.4 Fitness Assignment

It is important to note that domination rank $\mathcal{R}$ is a primary component of fitness assignment while sharing intensity $\mathcal{I}$ and crowding distance $\mathcal{D}$ prevent thickly crowded and dominated individuals from surviving through the selection process. Taking into account their effect on preference of individuals, a fitness function is defined as follows:

$$f(i) = \begin{cases} \mathcal{R}(i)\left(1 + \frac{1}{1+\mathcal{D}(i)}\right), & \text{if } i \in \mathcal{F}_1 \\ \mathcal{R}(i) + \gamma\mathcal{I}(i), & \text{otherwise.} \end{cases} \qquad (6.7)$$

Here, $f(i)$ denotes the fitness value of individual i, $\mathcal{R}(i)$ is the domination rank of individual i, γ is a regularization parameter for penalty, and $\mathcal{I}(i)$ and $\mathcal{D}(i)$ are the sharing intensity and the crowding distance of individual i, respectively. An individual with a lower fitness value is always preferable.

As the value of parameter γ increases, sharing effect on the fitness grows so that diversity is promoted. In the reverse case, proximity is emphasized due to the growth of the effect of domination rank on the fitness. A proper setting of γ may be crucial. Note that the value of the parameter γ should not be too large. This is because the fitness assignment gradually approaches that of NSGA-II as γ goes to 1. It is suggested that the parameter value not exceed the total number of selected individuals for adequately balancing between proximity and diversity.

An example of fitness assignment is depicted in Fig. 6.4. It is seen that an individual is penalized in proportion to the degree of domination and the extent of crowding. In other words, a lower fitness value is assigned to an individual that is closer to the first front $\mathcal{F}_1$ and less crowded while mostly representing population.

6.4.5 Elitism

Elitism allows the best solutions of the current generation to be copied into the next generation [32, 94] (see Sect. 4.2.2). Elitism plays an important role in MGEAs due to the availability of equally preferable multiple solutions [19,29,82,123]. It is important to retain many elitist solutions to further advance the set of nondominated solutions. In general, a variation operator such as recombination should be capable of generating new individuals that are better in the sense of proximity as well as diversity. Of course, the operator suffers as the set of nondominated solutions approaches the true Pareto front $\mathcal{F}^*$. If a nondominated solution gets lost in a certain generation, it is very hard to recover a new nondominated individual in its neighborhood. Thus, elitism directly contributes to *exploration* because it preserves superior individuals which are difficult to generate [19]. Moreover, elitism can also help in *exploitation* as it determines the individuals which are selected to survive the current generation [19]. Note that an easy but efficient approach for incorporating elitism into MGEAs is to employ *truncation selection*.

6.5 Real-valued Multiobjective Optimization Problems

This section presents real-valued multiobjective optimization problems (RMOPs): (additively) decomposable problems and some traditional benchmarks.

6.5.1 Decomposable Multiobjective Optimization Problems

Decomposable multiobjective optimization problems are created by combining basis functions of bounded difficulty. The value of a specific objective is computed by the sum of the corresponding objective values of all the basis functions.

The first problem is a multiobjective deceptive problem I (MDP-I). It is based on the binary MDP suggested in [59,60]. The MDP-I is composed of a real-valued deceptive problem (RDP) and its complement.

The RDP to be maximized is defined as follows:

$$f_{RDP}(\mathbf{y}) = \sum_{i=1}^{m} f_{trap}(y_{2i-1}, y_{2i}) \tag{6.8}$$

where $y_j \in [0, 1]$, $\forall j$, m are the number of subproblems, and the trap function f_{trap} which is the constituent of deceptive problems is given by

$$f_{trap}(y_j, y_{j+1}) = \begin{cases} \{\alpha/(1-\delta)\}\,(t_{j,j+1} - \delta)\,, & \text{if } t_{j,j+1} \geq \delta, \\ (\beta/\delta)\,(\delta - t_{j,j+1})\,, & \text{otherwise.} \end{cases} \quad (6.9)$$

Here, $t_{j,j+1}$ is given by $\sqrt{(y_j^2 + y_{j+1}^2)/2}$, α and β are the global and the local (i.e, deceptive) optimum, respectively, so that α/β indicates the signal to noise ratio (SNR), and δ is the border of attractors. Note that the function can be deceptive only if the constraint, $\alpha < (\alpha + \beta)\delta$, is satisfied.

It may be noted that the RDP has 2^m optimal solutions, among which there is only one global optimum. An unsavoury feature of RDP is that the optimum is isolated and there is a narrow basin of attraction toward the optimum. Thus, it is not readily amenable to hill-climbing strategy (such as mutation) alone. In order to efficiently solve the RDP, a recombination operator that can learn the linkage between variables and achieve the maximum BB-wise mixing rate is essential. This is because if the variables of each trap function are processed independently (as in uniform crossover), BBs (i.e., partial solutions) near to the basin of attraction toward the global optimum gradually degenerate, thereby eventually leading to one of the local optima.

On the other hand, the complement of RDP can be created by

$$\overline{f}_{RDP}(\mathbf{y}) = \sum_{i=1}^{m} \overline{f}_{trap}\,(y_{2i-1}, y_{2i}) \quad (6.10)$$

where the complement of trap function $\overline{f}_{trap}$ is defined by

$$\overline{f}_{trap}(y_j, y_{j+1}) = \begin{cases} \{\overline{\alpha}/(1-\delta)\}\,(t_{j,j+1} - \delta)\,, & \text{if } t_{j,j+1} \geq \delta, \\ (\overline{\beta}/\delta)\,(\delta - t_{j,j+1})\,, & \text{otherwise.} \end{cases} \quad (6.11)$$

Here, $\overline{\alpha}$ and $\overline{\beta}$ are the local and global optimum, respectively. There is no deception.

Note that the complementary RDP also has only one global optimum, but there is very wide attractor around it. Thus, any type of algorithm discovers the optimum. The problem itself is trivial, but it makes the deception of RDP harder when it is incorporated with the RDP in the context of multiobjective optimization. The reason is given below.

Let us consider the RDP and its complement simultaneously. The complement strongly leads the variables toward zero value where the global optimum of the complement occurs. Because of the deceptive characteristic of RDP, it is difficult to find the optimum that is encountered as all the variables reach "1". In other words, extracting and keeping the optimal BBs of the RDP is extremely hard due to the strong signal for the optimal solutions of the complementary RDP as well as the deception of the RDP itself. Thus, it is hard to discover the true Pareto front since it is composed of combinations of the optimal BBs of both the problems.

On the basis of the RDP and its complement, the MDP-I is defined as follows:

$$\text{maximize } f_{MDP\text{-}I} = (f_{RDP}(\mathbf{y}), \overline{f}_{RDP}(\mathbf{y})). \tag{6.12}$$

Main difficulties of the problem lie in deception, tight linkage, and multi-modality [27, 59, 60]. The true Pareto front has a total of $(m + 1)$ number of points. When they are listed from 0 to m along the objective values of the RDP, each ith point has $\binom{m}{i}$ distinct solutions [59, 60]. It is important to note that linkage-friendly recombination that can discover linear interactions between variables and then incorporate the knowledge to effectively inter-mix BBs is essential to successfully solve the MDP-I.

The second problem is a multiobjective deceptive problem II (MDP-II) that also consists of the RDP. The MDP-II is formulated as follows:

$$\begin{aligned}
\text{minimize } f_{MDP\text{-}II}(\mathbf{y}) &= (f_1(\mathbf{y}), f_2(\mathbf{y})), \\
\text{where } f_1(\mathbf{y}) &= y_1, \\
f_2(\mathbf{y}) &= g(\mathbf{y}) \left\{ 1 - (f_1(\mathbf{y})/g(\mathbf{y}))^2 \right\}, \\
g(\mathbf{y}) &= 1 + m - f_{RDP}(y_2, \cdots, y_n)
\end{aligned} \tag{6.13}$$

where $y_j \in [0, 1]$, $\forall j$, m is the number of subproblems, and n is the size of the problem. Since it involves all the characteristics of the RDP, the linkage-friendly recombination is necessary for finding the true Pareto front. Note that there are $(m + 1)$ classes of Pareto fronts. As in the case of the MDP-I, each ith class has $\binom{m}{i}$ distinct Pareto fronts as arranged by the objective values of the RDP.

The last decomposable problem is a multiobjective nonlinear, symmetric problem (MNSP). The problem employs Rosenbrock function as a basis function of a real-valued nonlinear, symmetric problem (RNSP). The RNSP is formulated as follows:

$$f_{RNSP}(\mathbf{y}) = \sum_{i=1}^{m} f_{Rosen}(y_{(i-1) \cdot k + 1}, \cdots, y_{i \cdot k}) \tag{6.14}$$

where $y_j \in [-5.12, 5.12]$, $\forall i$, k and m is the subproblem size and the number of subproblems, and the Rosenbrock function f_{Rosen} is given by

$$f_{Rosen}(y_1, \cdots, y_k) = \sum_{i=2}^{k} \{100 \cdot (y_i - y_{i-1}^2)^2 + (1 - y_{i-1})^2\}. \tag{6.15}$$

The RNSP is unimodal even though it consists of a number of flat curved valleys. Thus, if selection pressure is not very high, any algorithm may gradually transit to the global optimum without paying any special attention to learning and exploiting linkage information. However, it leads to quite a slow convergence as well as huge population size, and worse still, this problem is proportional to the problem dimensionality. Consequently, it is quite hard to find the global optimum unless linkage-friendly recombination is incorporated.

By incorporating the RNSP, the MNSP is created as follows:

$$\text{minimize } f_{MNSP}(\mathbf{y}) = (f_1(\mathbf{y}), f_2(\mathbf{y})),$$
$$\text{where } f_1(\mathbf{y}) = y_1,$$
$$f_2(\mathbf{y}) = g(\mathbf{y})\left\{1 - \sqrt{f_1(\mathbf{y})/g(\mathbf{y})}\right\},$$
$$g(\mathbf{y}) = 1 + f_{RNSP}(y_2, \cdots, y_n) \tag{6.16}$$

where $y_1 \in [0, 1]$ and $y_j \in [-5.12, 5.12]$ for $2 \le j \le n$, and n is the size of the problem.

Note that linkage-friendly recombination that is capable of capturing and incorporating nonlinear, symmetric interactions is required for efficiently solving the MNSP. In this regard, it ensures that the MNSP offers a real challenge for real-coded multiobjective optimization algorithms, especially EDA-based algorithms, due to some difficulties arising from nonlinear, symmetric linkage between variables and flat, long basin of attraction of each subproblem.

6.5.2 Traditional Multiobjective Optimization Problems

Many RMOPs have been proposed with a view to testifying whether multiobjective optimization algorithms have the capability of dealing with a variety of difficulties. Three well-known difficult RMOPs are investigated here. They have some intriguing characteristics (beyond decomposability) which most multiobjective optimization algorithms find hard to negotiate.

The first test problem is the ZDT$_4$ function [122]. It is formulated as follows:

$$\text{minimize } f_{ZDT_4}(\mathbf{y}) = (f_1(\mathbf{y}), f_2(\mathbf{y})),$$
$$\text{where } f_1(\mathbf{y}) = y_1,$$
$$f_2(\mathbf{y}) = g(\mathbf{y})\left\{1 - \sqrt{f_1(\mathbf{y})/g(\mathbf{y})}\right\},$$
$$g(\mathbf{y}) = 1 + 10(n-1) + \sum_{j=2}^{n}\left\{y_j^2 - 10cos(4\pi y_j)\right\} \tag{6.17}$$

where $y_1 \in [0, 1]$ and $y_j \in [-5, 5]$ for $2 \le j \le n$, and n is the size of the problem. It has a very large number, viz., 21^9, of local Pareto fronts [29, 122]. Furthermore, the number of Pareto fronts increases as the individuals approach the true Pareto front. Due to the high multimodality, it is very hard to discover the true Pareto front.

The second test problem is the ZDT$_6$ function [122]. It is defined as follows:

$$\text{minimize } f_{ZDT_6}(\mathbf{y}) = (f_1(\mathbf{y}), f_2(\mathbf{y})),$$
$$\text{where } f_1(\mathbf{y}) = 1 - e^{-4y_1}sin^6(6\pi y_1),$$
$$f_2(\mathbf{y}) = g(\mathbf{y})\left\{1 - (f_1(\mathbf{y})/g(\mathbf{y}))^2\right\},$$
$$g(\mathbf{y}) = 1 + 9\left\{\sum_{i=2}^{n} y_j/(n-1)\right\}^{1/4} \tag{6.18}$$

where $y_j \in [0,1]$, $\forall j$, and n is the size of the problem. It has global Pareto-optimal solutions which are nonuniformly distributed along the true Pareto front such that more solutions come out as $f_1(\mathbf{y})$ goes to 1. Moreover, the density of the solutions increases away from the true Pareto front. Thus, it is quite difficult to achieve a good spread of solutions along the true Pareto front.

The last test problem is a modified BT_1 (mBT_1) function. With a view to emphasizing interactions of variables, the original BT_1 proposed in [20] has been amended as follows:

$$\text{minimize } f_{mBT_1}(\mathbf{y}) = (f_1(\mathbf{y}), f_2(\mathbf{y})),$$
$$\text{where } f_1(\mathbf{y}) = y_1,$$

$$f_2(\mathbf{y}) = 1 - f_1(\mathbf{y}) + \sum_{j=1}^{n} \left| y_j + \sum_{i=1}^{j-1} y_i \right|$$

where $y_1 \in [0,1]$ and $y_j \in [-3,3]$ for $2 \le j \le n$, and n is the size of the problem. Unlike the previous two functions, it has multivariate linear dependencies between variables. Hence, discovering the true Pareto front is not easy without incorporating the knowledge on the problem structure.

6.6 Experimental Results and Discussion

This section describes the performance metrics and the experimental setup. The performance of MrBOA on RMOPs under various conditions is also investigated.

6.6.1 Performance Measures

The main goal in multiobjective optimization is to achieve higher proximity of the nondominated set of solutions while preserving better diversity. Hence, proper metrics are required for efficiently assessing the performance of an algorithm regarding proximity and diversity. In this regard, we employ the following two metrics.

The first is the *proximity* metric [20,117]. It measures the extent of convergence of the nondominated set to the true Pareto front. The proximity metric is given by,

$$\Upsilon = \frac{1}{|\mathcal{F}_1|} \sum_{\mathbf{z}^0 \in \mathcal{F}_1} \min_{\mathbf{z}^1 \in \mathcal{F}^*} \{d_E(\mathbf{z}^0, \mathbf{z}^1)\} \tag{6.19}$$

where $d_E(\mathbf{z}^0, \mathbf{z}^1)$ is the Euclidean distance between objective values and it is given by

$$d_E(\mathbf{z}^0, \mathbf{z}^1) = \sqrt{\sum_{k=1}^{M} (f_k(\mathbf{y}^1) - f_k(\mathbf{y}^0))^2} \tag{6.20}$$

where M is the number of objectives. A smaller value denotes a higher proximity of the nondominated set of solutions.

The second is the *diversity* metric. It measures the extent of spread of the nondominated solutions. The diversity metric is defined by

$$
\Delta = \frac{d_f + d_l + \sum_{i=1}^{|\mathcal{F}_1|-1} |d_i - \mu_d|}{(|\mathcal{F}_1| - 1)\mu_d}
\tag{6.21}
$$

where d_i and μ_d are the Euclidean distance between consecutive solutions in the set $\mathcal{F}_1$ and the average of these distances respectively, and d_f and d_l are the Euclidean distances between the extreme solutions and the boundary solutions of the computed nondominated set. The concrete methodology of computing d_f and d_l can be found in [29]. A smaller value indicates a better diversity of the nondominated solutions.

Note that the diversity metric is essentially identical to that of NSGA-II [29], but the difference lies in the consistency in indicating good/bad distribution of the nondominated solutions. With regard to diversity, the worst case occurs when all the solutions come together to form a single solution. Moreover, the diversity improves as the variance of d_i and the values of d_f and d_l decrease. Thus, the diversity metric always has a small value for well distributed nondominated set of solutions. It goes to infinity as they gather together.

Note that an effective comparative study of the two algorithms under consideration must be performed by properly treating proximity and diversity. However, there is a simple but effective alternative for directly comparing the algorithms even though it does not provide any information about proximity and diversity of nondominated solutions. This is the *coverage* metric [122,123]. It is used for comparing the dominance relationship between two sets of nondominated solutions resulting from two different algorithms. We employ a modified coverage metric [34] that compensates some undesirable mathematical properties (as a metric). It is defined as follows:

$$
\mathcal{C}(\mathcal{Q}_1^A, \mathcal{Q}_1^B) = \frac{|\{\mathbf{y}^1 \in \mathcal{Q}_1^B | \exists \mathbf{y}^0 \in \mathcal{Q}_1^A : \mathbf{y}^0 \succ \mathbf{y}^1\}|}{|\mathcal{Q}_1^B|}
\tag{6.22}
$$

where $\mathcal{Q}_1^A$ ($\mathcal{Q}_1^B$) presents the set of solutions in the first front $\mathcal{F}_1$ (i.e., nondominated solutions) returned by the algorithm A (B). The metric measures the proportion of the members of $\mathcal{Q}_1^B$ that are strictly dominated by the members of $\mathcal{Q}_1^A$. Hence, $\mathcal{C}(\mathcal{Q}_1^A, \mathcal{Q}_1^B) = 1$ if all the individuals in $\mathcal{Q}_1^B$ are dominated by those in $\mathcal{Q}_1^A$; on the contrary, $\mathcal{C}(\mathcal{Q}_1^A, \mathcal{Q}_1^B) = 0$ in the opposite case. Moreover, $\mathcal{C}(\mathcal{Q}_1^A, \mathcal{Q}_1^A) = 0$, and $\mathcal{C}(\mathcal{Q}_1^A, \mathcal{Q}_1^B) = 0$ if $\mathcal{Q}_1^A$ and $\mathcal{Q}_1^B$ are subsets of a nondominated set. Note that both $\mathcal{C}(\mathcal{Q}_1^A, \mathcal{Q}_1^B)$ and $\mathcal{C}(\mathcal{Q}_1^B, \mathcal{Q}_1^A)$ must be considered independently since they have physically different meanings.

Although the above performance metrics are very rational for investigating performance of multiobjective optimization algorithms on most of MOPs, they are not suitable for applying to the MDP-I. The reason is given below.

The true Pareto front of MDP-I is composed of several distinct points. The goal is to find all solutions corresponding to the true Pareto front. But the deceptive property prohibits individuals from evolving toward optimal BBs. When an algorithm is completely under the influence of deception, all the individuals converge to one of the points that is only composed of the optimal BBs of the complementary RDP so that the proximity metric produces zero value, viz., a perfect proximity. This is no meaningful situation. We can obtain a similar result for the coverage metric. Hence, another metric must be considered. In some past studies [19, 20], a metric has been proposed for reflecting both proximity and diversity by computing the average distance from the true Pareto front to nondominated set. In this work, it is denoted as *prox-div* (PD) metric.

The PD metric is defined by

$$\Psi = \frac{1}{|\mathcal{F}^*|} \sum_{\mathbf{z}^1 \in \mathcal{F}^*} \min_{\mathbf{z}^0 \in \mathcal{F}_1} \{d_E(\mathbf{z}^0, \mathbf{z}^1)\}. \tag{6.23}$$

In the MDP-I, it returns zero value when all the points in the true Pareto front are found. If individuals converge to some of them, it yields a higher value. That is, the deception nudges the metric to a larger value. Moreover, a small value is obtained as the algorithm keeps the individuals which are approaching toward all the Pareto points. Note that the PD metric is conceptually akin to the solution quality in the single objective optimization.

6.6.2 Experiment Setup

The performance of MrBOA is compared with that of NSGA-II [29] and MIDEA [18, 19]. NSGA-II and MIDEA have been perceived to be representative algorithms of *population-based* (especially *advanced*) and *EDA-based* approaches, respectively. To assess the performance of algorithms, PD metric is used for MDP-I, and proximity, diversity and coverage metrics are employed for MDP-II, MNSP, ZDT$_4$, ZDT$_6$, and mBT$_1$.[4]

The NSGA-II employs real-coded encoding scheme with simulated binary crossover (SBX) operator, and polynomial mutation [28, 29]. Moreover, it uses a crossover probability of $p_c = 0.9$, crossover distribution index of $\eta_c = 20$, mutation probability of $p_m = 1/n$ (where n is the problem size), and a mutation distribution of $\eta_m = 20$. Although the parameter setting may not be the best, it has been exploited by the original study to achieve better overall performance.

The MrBOA uses normal mixture distributions obtained by clustering the selected individuals in order to learn probabilistic models. As a computationally efficient clustering mechanism, the K-means algorithm is employed for model selection and the randomized leader algorithm (RLA) with a threshold

---

[4] The reason has been described in Sect. 6.6.1.

value of 0.3 is used for model fitting. Truncation selection with parameter $\tau = 0.5$ and the Bayesian information criterion (BIC) with regularization parameter $\lambda = 0.5$ are invoked for learning a probabilistic model. As an elite-preserving strategy, the worst half of the population is replaced by the newly generated individuals. The number of allowable parents is given by half the problem size, viz., $\lfloor 0.5n \rfloor$. Moreover, the penalty parameter γ in the fitness assignment is set to 20. The set of parameter values has been determined on the basis of empirical investigation.

The MIDEA contains, in part, the same parameters that MrBOA retains due to a similarity of their basis algorithms (i.e., IDEA and rBOA). In the interest of fair comparison, the parameters – viz., the threshold value of the leader algorithm, the parameter τ of the truncation selection, the regularization parameter λ of BIC, and the number of maximum parents – have the same values as in MrBOA. Diversity preservation parameter that adjusts the size of preselection is set to 1.5 in the diversity-preserving selection.

The population size used is empirically obtained for each problem to get a good performance within the maximum number of (multiobjective) function evaluations. Here, MrBOA plays the role of a decision maker. The computed population size is assigned to the references. However, it may be inadequate for NSGA-II[5] because EDAs generally requires a larger population than does the population-based GEAs [89, 91]. In other words, there exists a different optimal population for each algorithm. However, if a large enough population is considered with sufficient number of generations to achieve acceptable convergence, this inadequacy would vanish. In the problems under focus, it has been empirically observed that a maximum of $50 \cdot 10^3$ evaluations is adequate for satisfactory convergence of all the algorithms. In this case, adequate population size N amounts to 1000 for MDP-I, 800 for MDP-II and ZDT_4, and 400 for MNSP, ZDT_6 and mBT_1.

The Pareto fronts presented are obtained by nondominated solutions of the set of nondominated solutions collected from 30 runs. At the last, all the results are averaged over all the runs.

6.6.3 Results and Discussion

Figure 6.5 compares the Pareto fronts found by the algorithms as applied to MDP-I. The parameter values are: $\alpha = \overline{\beta} = 1.0$, $\beta = \overline{\alpha} = 0.8$, $\delta = 0.8$, and $m = 5$. The figure shows that MrBOA discovers the true Pareto front, while NSGA-II finds only two most deceptive points. That is, NSGA-II does not cope with deception. In this respect, MIDEA is superior to the NSGA-II because it can discover most of the points in the true Pareto front although the speed of convergence seems to be lower than those of the others. However,

---

[5] In this study, the population size of NSGA-II denotes the complete population size, viz., a sum of parents and offsprings, for notational consistency with MIDEA and MrBOA.

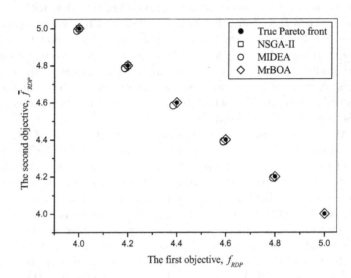

Fig. 6.5. Pareto fronts obtained from algorithms on $f_{MDP\text{-}I}$.

Table 6.1. Statistical comparison of algorithms on $f_{MDP\text{-}I}$.

Measure	NSGA-II	MIDEA	MrBOA
μ_Ψ	0.6804	0.2848	0.0304
σ_Ψ	0.0697	0.0818	0.0316
Statistical t-test			
Measure	NSGA-II − MIDEA	NSGA-II − MrBOA	MIDEA − MrBOA
t-value	21.44†	49.34†	18.44†
Order	MrBOA $\succ$ MIDEA $\succ$ NSGA-II		

† The value of t is *significant* at $\alpha = 0.01$ by a paired, two-tailed test. The symbols $\succ$ and $\sim$ represent *dominance* and *indifference* between algorithms.

the MIDEA never finds all BBs of the RDP. That is, the MIDEA can deal with deception to some extent but it is also not free form the handicap. Table 6.1 supports the claim that the MrBOA mostly finds the true Pareto front while the MIDEA and the NSGA-II do not return good performance.

Figure 6.6 depicts the Pareto fronts for MDP-II with $m = 5$ and $n = 11$. In the figure, solid lines present all kinds of Pareto fronts whose degree of deception increases outwards. It is seen that MrBOA converges to the true Pareto front while MIDEA and NSGA-II do not do so. Further, the NSGA-II finds the solutions that are close to the most deceptive front: it converges to the second most deceptive front. It seems that all the algorithms maintain relatively good spread of their nondominated solutions although MIDEA ex-

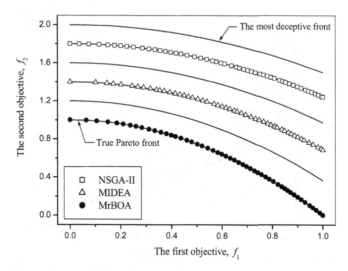

Fig. 6.6. Comparison results of the Pareto fronts for f_{MDP-II}.

poses some wider gap between solutions. It follows that MrBOA achieves a higher proximity than do the references; and all the algorithms appear to have similar diversities (MIDEA might be worse than the others, however). In this regard, more details are provided in Table 6.2. The table supports the claim on the predominance of MrBOA over NSGA-II and MIDEA with regard to proximity as well as diversity.

Figure 6.7 compares the results for MNSP with $k = 3$, $m = 3$, and $n = 10$. As a reference for a more precise assessment, an upper curve obtained when $f_{RNSP} = 0.1$ whose solution is far away from the optimum is presented in the figure. As seen in the figure, the MrBOA offers near-optimal and uniformly distributed Pareto front, while MIDEA show the worst performance for both proximity and diversity measures. Although NSGA-II seems, on the face of it, to return a good Pareto front, the quality is clearly very poor as compared with the reference front. Table 6.2 also upholds the superiority of MrBOA.

From the above results, we may conclude that the MrBOA outperforms the NSGA-II and the MIDEA for decomposable problems with some difficulty such as deception or nonlinearity with regard to both proximity and diversity. Note that the superiority of MrBOA obtains from its ability to discover problem regularities and inter-mixing superior partial solutions by referring to the accumulated information.

Figure 6.8 demonstrates the nondominated solutions (in the objective space) found by each algorithm when applied to ZDT_4 with $n = 10$. The results of proximity and diversity are quantified in Table 6.2. It is seen that

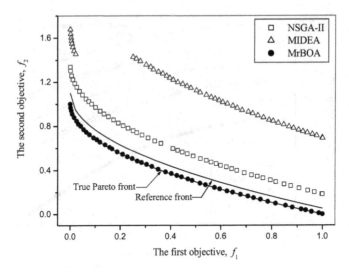

Fig. 6.7. Computed Pareto fronts with algorithms on f_{MNSP}.

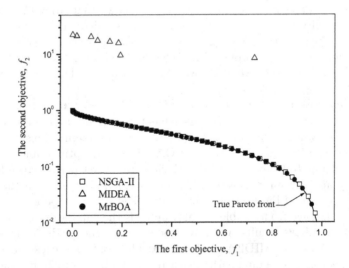

Fig. 6.8. Comparison results of the Pareto fronts for f_{ZDT_4}.

NSGA-II and MrBOA[6] tend to converge to the true Pareto front. Although the NSGA-II achieves a better diversity than the MrBOA, there is no differ-

---

[6] Due to high multimodality feature of ZDT$_4$, the threshold value of 0.05 for the RLA is used.

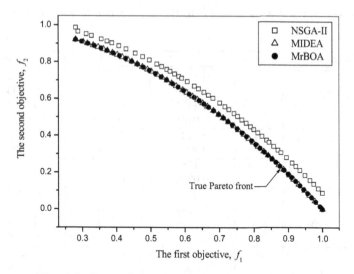

Fig. 6.9. Pareto fronts found by algorithms for f_{ZDT_6}.

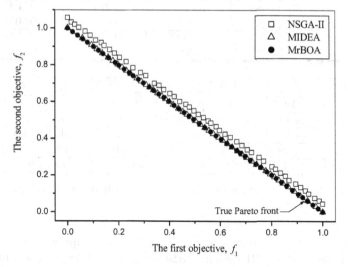

Fig. 6.10. Nondominated solutions returned by algorithms for f_{mBT_1}.

ence in their proximity values. The results clearly validate the inferiority of MIDEA. That is, the MIDEA cannot adequately deal with multimodality.

Table 6.2. Statistical comparison of algorithms on $f_{MDP\text{-}II}$, f_{MNSP}, f_{ZDT_4}, f_{ZDT_6} and f_{mBT_1}.

Problem	Measure	NSGA-II	MIDEA	MrBOA
$f_{MDP\text{-}II}$ $(m = 5, n = 11)$	μ_Υ	0.9399	0.6511	0.1172
	σ_Υ	0.0360	0.1541	0.1193
	μ_Δ	0.2455	0.3269	0.2049
	σ_Δ	0.0151	0.1601	0.0147
f_{MNSP} $(m = 3, n = 10)$	μ_Υ	0.7221	1.2775	0.1270
	σ_Υ	0.3954	0.5748	0.1602
	μ_Δ	0.1786	0.2629	0.1634
	σ_Δ	0.0242	0.0384	0.0117
f_{ZDT_4} $(n = 10)$	μ_Υ	0.0027	23.252	0.0252
	σ_Υ	9.5e-4	2.9381	0.0601
	μ_Δ	0.1829	0.5535	0.3178
	σ_Δ	0.0406	0.2291	0.1676
f_{ZDT_6} $(n = 10)$	μ_Υ	0.0737	0.0444	0.0213
	σ_Υ	0.0097	0.0527	0.0248
	μ_Δ	0.3089	0.6496	0.7682
	σ_Δ	0.0198	0.5274	0.5519
f_{mBT_1} $(n = 10)$	μ_Υ	0.1000	0.0245	0.0034
	σ_Υ	0.0691	0.0531	0.0032
	μ_Δ	0.5930	0.4108	0.2751
	σ_Δ	0.2869	0.0760	0.0214

Statistical t-test; (Υ, Δ)			
Problem	NSGA-II − MIDEA	NSGA-II − MrBOA	MIDEA − MrBOA
$f_{MDP\text{-}II}$	$(10.26^\dagger, -2.82^\dagger)$	$(38.08^\dagger, 9.86^\dagger)$	$(14.30^\dagger, 4.10^\dagger)$
f_{MNSP}	$(-4.58^\dagger, -11.99^\dagger)$	$(7.93^\dagger, 2.76^\dagger)$	$(10.96^\dagger, 14.14^\dagger)$
f_{ZDT_4}	$(-43.34^\dagger, -8.77^\dagger)$	$(-2.07, -4.15^\dagger)$	$(43.38^\dagger, 3.65^\dagger)$
f_{ZDT_6}	$(2.93^\dagger, -4.49^\dagger)$	$(10.24^\dagger, -4.53^\dagger)$	$(2.04, -0.13)$
f_{mBT_1}	$(4.37^\dagger, 3.52^\dagger)$	$(7.64^\dagger, 6.08^\dagger)$	$(2.15, 9.46^\dagger)$

† The value of t is *significant* at $\alpha = 0.01$ by a paired, two-tailed test.

Figure 6.9 shows the results of the algorithms as applied to ZDT$_6$ with $n = 10$. It seems that MIDEA and MrBOA[7] find the Pareto fronts quite close to the true front while NSGA-II returns the nondominated solutions a little away from the global optimum. Further information on their performances is also presented in Table 6.2. As seen in the table, there is no dominance among the algorithms with regard to proximity, while the diversity of NSGA-II and MrBOA is superior to that of MIDEA. Moreover, high deviation in MIDEA's performances signals its instability.

---

[7] With a view to enduring the nonuniformity of ZDT$_6$, one component is employed for model selection and model fitting; $K = 1$ and *threshold* $= 1.0$ for the K-means algorithm and the RLA, respectively.

Table 6.3. Dominance comparison of algorithms based on proximity and diversity.

Problem	Proximity; Υ	Diversity; Δ
$f_{MDP\text{-}II}$	MrBOA $\succ$ MIDEA $\succ$ NSGA-II	MrBOA $\succ$ NSGA-II $\succ$ MIDEA
f_{MNSP}	MrBOA $\succ$ NSGA-II $\succ$ MIDEA	MrBOA $\succ$ NSGA-II $\succ$ MIDEA
f_{ZDT_4}	MrBOA $\sim$ NSGA-II $\succ$ MIDEA	NSGA-II $\succ$ MrBOA $\succ$ MIDEA
f_{ZDT_6}	MrBOA $\sim$ MIDEA $\succ$ NSGA-II	NSGA-II $\succ$ MrBOA $\sim$ MIDEA
f_{mBT_1}	MrBOA $\sim$ MIDEA $\succ$ NSGA-II	MrBOA $\succ$ MIDEA $\succ$ NSGA-II

The symbols $\succ$ and $\sim$ represent *dominance* and *indifference* between algorithms.

Figure 6.10 compares the Pareto fronts for mBT$_1$ with $n = 10$. Moreover, Table 6.2 provides more informative results. It is observed that the nondominated solutions of MIDEA and MrBOA nearly converge on the optimal front but NSGA-II is misled by some local front. In detail, proximity performance of MrBOA is comparable to that of MIDEA, and both the results are superior to that of NSGA-II. With regard to diversity performance, MrBOA is the best and NSGA-II is the worst.

From the results for ZDT$_4$, ZDT$_6$ and mBT$_1$, it may be concluded that MrBOA finds the set of nondominated solutions whose proximity is comparable (or even better) to that of NSGA-II or MIDEA for the problems with some difficulty over decomposability, without compromising on diversity of the solutions.

Moreover, Table 6.3 compares dominance relation among the algorithms on the basis of proximity and diversity measures given in Table 6.2. In the comparative study, proximity is a matter of primary importance. The table indicates that MrBOA achieves the best performance regardless of problem difficulties in respect of decomposability (interlaced with deception or nonlinearity), multimodality, nonuniformity and variables' interactions; while MIDEA is superior to NSGA-II except for MNSP and ZDT$_4$ which have some difficulties with nonlinearity under decomposability and multimodality, respectively.

On the other hand, simple direct comparison with regard to the coverage metric is performed in Fig. 6.11. A *box plot* has a notched box summarizing 50% of the data, viz., the lower and upper boundaries (of the box) are the lower and upper quartiles. The appendages indicate 10% and 90% percentiles respectively, while the remainders represent outliers. The square symbol presents the mean value. In order to understand the correct relationships, it is necessary to take both side coverage metrics into consideration. With the same notations as in Sect. 6.6.1, the way of interpretation is described as follows. When $\mathcal{C}(\mathcal{Q}_1^A, \mathcal{Q}_1^B) = 1$ and $\mathcal{C}(\mathcal{Q}_1^B, \mathcal{Q}_1^A) = 0$, the algorithm A always dominates the algorithm B, and vice versa. If $\mathcal{C}(\mathcal{Q}_1^A, \mathcal{Q}_1^B)$ and $\mathcal{C}(\mathcal{Q}_1^B, \mathcal{Q}_1^A)$ have 0, they do not dispute preeminence with each other. Further, it is impossible that both the metrics have 1.

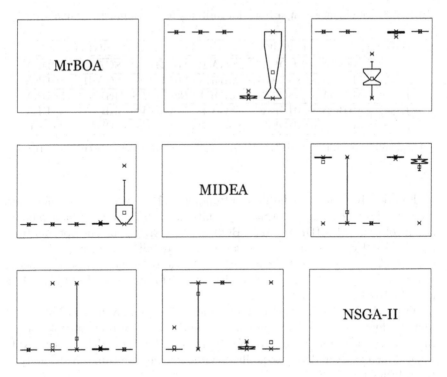

Fig. 6.11. Comparison of dominance relationships based on coverage metric. The box plots arrange the results for $f_{MDP\text{-}II}$, f_{MNSP}, f_{ZDT_4}, f_{ZDT_6} and f_{mBT_1} from the left to the right. The scale is 0 at the bottom and 1 at the top. Each rectangle represents the coverage metrics for algorithm A of the corresponding row and algorithm B of the corresponding column, i.e., $\mathcal{C}(A, B)$.

The figure shows that the Pareto fronts reached by NSGA-II and MIDEA for the MDP-II and the MNSP are entirely covered by those of MrBOA. In the ZDT$_4$, the MrBOA discover the nondominated solutions that always dominate the solution set of the MIDEA and has a relatively low probability of being dominated by the solutions of NSGA-II. In the ZDT$_6$, the MIDEA nearly bears comparison with the MrBOA, while the Pareto front of NSGA-II is almost behind that of MrBOA. In the case of mBT$_1$, MrBOA can dominate as well as be dominated by MIDEA with a small chance (i.e., less than 0.4), but it strictly dominates NSGA-II. The results are generally consistent with those obtained by considering proximity and diversity metric which offer more accurate comparison (see Table 6.3). Two algorithms may not be statistically different even though the coverage metric strictly exhibits difference between their nondominated sets. Although any information about the extent of closeness and spread of nondominated solutions cannot be obtained, the metric is definitely useful for conveniently testing dominance relationship between algorithms.

In conclusion, the proposed MrBOA achieves higher proximity and better diversity of nondominated solutions for the problems of bounded difficulty while finding comparable or better solutions to the problems with some difficulty beyond decomposability.

6.7 Summary

In this chapter, we have developed a multiobjective real-coded Bayesian optimization algorithm (MrBOA) as an efficient tool for dealing with various difficult multiobjective optimization problems (MOPs), especially decomposable ones. It was achieved by fitting the competent (single-objective) real-coded Bayesian optimization algorithm (rBOA) into multiobjective optimization framework, without weakening its ability to automatically discover problem regularities and thoroughly exchange superior characters of current solutions. To further preserve diversity of population without degrading proximity, a selection method has been developed by combining Pareto ranking with the proposed diversity-preserving mechanisms – adaptive sharing and dynamic crowding. The adaptive sharing tries to accord preference to less dominated solutions, and the dynamic crowding offers a higher chance of survival to sparse nondominated solutions. The selection scheme is characterized by a unique fitness assignment scheme. Individual's fitness is assigned by incorporating the domination rank with some penalty imposed on the sharing and the crowding schemes.

Experimental studies demonstrated that MrBOA definitely achieves statistically competitive results with the two state-of the art MGEAs – NSGA-II and MIDEA – on deceptive or nonlinear-symmetric decomposable problems in finding a near-optimal and uniformly distributed Pareto front. For some traditional MOPs with multimodality, nonuniformity of the Pareto front, or close interactions of variables, the MrBOA was able to find a nondominated set whose proximity and diversity are comparable to or even better than NSGA-II and MIDEA. It is noted that the competency of MrBOA is largely brought about by the unique strengths of rBOA; and the proposed selection method also plays a role in further enhancing the performance.

The proposed MrBOA is thought to be an effective tool for dealing with many engineering and scientific problems whose difficulties might be beyond the reach of current techniques.

7

Conclusions

This chapter summarizes and concludes this book. Also, directions for future research have been indicated.

7.1 Summary

Primary contributions of this book are summarized below.

- **Design of practical genetic algorithms:** The book offered practical design guidelines for developing efficient genetic algorithms (GAs) to successfully solve real-world problems. As a critical design component, a practical population-sizing model was also developed, which accurately estimates an adequate population size with a desired quality of solution without requiring statistical knowledge about problems such as the signal and variance of competing building blocks (BBs). Effectiveness of the model was supported by the results on test problems.
- **Application to real-world problems:** The book developed a GA that efficiently solves an important real-world application – shortest path routing problem. Operators were designed by following the suggested road map. The algorithm can traverse the search space effectively and speedily without overly depending on problem types. Experimental results exhibited that the proposed algorithm outperforms many competitors. The results also emphasized the relevance of the population-sizing model in a practical setting.
- **Elitist compact genetic algorithms:** Two elitism-based compact genetic algorithms (cGAs) were proposed in a simple estimation of distribution algorithm (EDA) framework: the persistent elitist cGA (pe-cGA) and the nonpersistent elitist cGA (ne-cGA). The pe-cGA counters selection noise by keeping the current best solution until, hopefully a better solution is found. It was shown that it incorporates a model that is equivalent

Chang Wook Ahn: *Advances in Evolutionary Algorithms: Theory, Design and Practice*, Studies in Computational Intelligence (SCI) **18**, 153–157 (2006)
www.springerlink.com   © Springer-Verlag Berlin Heidelberg 2006

to (1+1) evolutionary strategy (ES) with self-adaptive mutation. The ne-cGA further enhances the performance of the pe-cGA by avoiding strong elitism (i.e., high selection pressure) that may lead to premature convergence. The improvement is a consequence of the fact that it does not allow genetic diversity to degenerate. It was also shown that the allowable scope of an elite member's inheritance is bounded by the simulated population size. Furthermore, an analytic speedup model that quantifies convergence improvement was developed. Experimental studies demonstrated the superiority of the proposed algorithms with regard to solution quality as well as convergence speed. The results also pointed to the correctness of the speedup model.

- **Real-coded Bayesian optimization algorithm (rBOA):** A competent real-coded EDA, viz., rBOA, has been developed for solving a wide class of real-world problems efficiently, quickly, and reliably. Problem substructures were extracted from the Bayesian factorization graph involving finite mixture models. Each substructure was fitted by mixture distributions, and new candidate solutions were generated by an independent subproblem-wise sampling procedure. The power of rBOA arises from its ability to effect proper decomposition and probabilistic building-block crossover (PBBC) on real-valued multivariate data (i.e., a class of mechanisms for discovering and exploiting problem regularities). A scalability model of the rBOA was also developed on problems of bounded difficulty by computing the number of fitness evaluations until convergence to the optimal solution. The theory asserted that the rBOA finds the optimal solution with a sub-quadratic scale-up behavior in respect of the problem size. Experimental studies showed that the rBOA outperforms advanced real-coded EDAs when faced with decomposable problems regardless of inherent problem characteristics. For nondecomposable problems, comparable or better performance was achieved. The experiments also supported the analytic model on sub-quadratic scalability of rBOA.

- **Multiobjective rBOA (MrBOA):** The book devised a competent multiobjective EDA (MEDA) , viz., MrBOA, for solving difficult multiobjective optimization problems (MOPs) of bounded difficulty. It was achieved by extending the proposed (single-objective) rBOA so as to allow for automatic discovery and effective exploitation of regularities in MOPs. The unique characteristics of the rBOA – proper decomposition and PBBC – invest the MrBOA with appreciable power. A selection method was developed for preserving diversity. The proximity to the Pareto front was not compromised as a result. The selection is characterized by a unique fitness assignment scheme that incorporates domination rank with some penalty imposed on sharing intensity and crowding distance. Experimental results demonstrated that the MrBOA achieves a better performance for decomposable and nondecomposable MOPs than do state-of-the-art multiobjective GEAs (MGEAs).

7.2 Future Work

There is a bright future for research on rBOA and MrBOA. In this book, we have focused on the development of a class of procedures that performs proper decomposition and effective search through intermixing superior partial solutions (i.e., BBs). Further, rBOA and MrBOA have been developed for largely dealing with problems with single-level decomposition. Thus, several paths for future work are open.

7.2.1 Incorporating Efficiency-Enhancement Techniques

There are promising approaches to enhancing the efficiency of rBOA and MrBOA. *Parallelization, hybridization, evaluation relaxation*, and *incremental and sporadic model building* are representative techniques [24, 60, 89].

Parallelization can considerably reduce execution time and possibly find a better solution by implementing an algorithm over multiple processors. Parallelization can be realized by distributing fitness evaluation, model building and sampling, or the population [22, 89]. Several approaches have been investigated in this regard [22, 64, 82]. Parallelization of model building and sampling may be a promising approach for rBOA and MrBOA because dealing with probabilistic models is generally computationally very costly. Of course, a combined approach may have some advantages. There is another approach under the EDA framework [5]. The approach tries to conceptually combine two parallelization methods of decentralizing model building/sampling and distributing the population (although more work needs to be done).

Hybridization can improve performance of an algorithm by combing the advantages of the algorithm and other types of competent algorithms [15, 89, 107]. A well-known approach is to properly balance between global and local search algorithms. These kinds of schemes are generally known as *memetic* algorithms. The hybridization limits the search to local optima, thereby allowing problem regularities to be easily identified. The required population size can be significantly shrunk in this way. When incorporating local search into rBOA and MrBOA, intensive uses of local search in the early and final stages can help improve the initial probabilistic model by rapidly capturing promising regions and enhance convergence performance by dense-searching in the vicinity of the global optimum, respectively.

Evaluation relaxation offers an effective tool for evaluating a large number of candidate solutions quickly and reliably [24, 89, 102]. It can be achieved by approximating/estimating fitness of candidate solutions from actual fitness of partial candidates. Although there is discrepancy between true and approximate fitness, the approximation error decreases as generations pass so that the global optimum can be discovered. The approach is quite effective when fitness evaluation is computationally expensive. The approach can significantly improve the performance of rBOA and MrBOA although more work related to the effect of the approximation error on model building is necessary.

Incremental and sporadic model building can significantly reduce the computational cost of learning the structure of a model [33, 89]. Since rBOA and MrBOA construct the model from the bottom in every generation, the approach can considerably enhance their efficiency. If a bad model is built in a specific generation, the incorrect information propagates into subsequent generations so that convergence performance is somewhat compromised. However, it does not endanger convergence to the optimum [89].

7.2.2 Challenging to Hierarchical Difficulty

The rBOA and MrBOA mainly target (single-objective and multiobjective) optimization problems that can be decomposed into subproblems of bounded order. There are many complex problems that can be decomposed into a hierarchy of levels of difficulty rather than into a single level [89, 93]. Moreover, the problems may have exponentially many local optima that hinder any local search from approaching the global optimum.

Hierarchical decomposition is required for dealing with hierarchical problems quickly, accurately, and reliably. To realize hierarchical decomposition, there are three key components – *decomposition, chunking,* and *niching* [89]. In this regard, rBOA and MrBOA can be readily extended because they already contain major components that lead to hierarchical rBOA (hrBOA) and multiobjective hrBOA (MhrBOA). This is described below.

In rBOA and MrBOA, a proper decomposition is ensured by constructing probabilistic models to discover important problem regularities. Sampling those models generates new candidates. Modeling multivariate data by incorporating finite mixture models can be viewed as an instance of chunking in the sense that all the parameters that encode all the explicit information about the data are compressed by several parameters that proportionally encodes alternative partial solutions to the particular subproblem. However, more work on grouping of decision variables from each subproblem of the lower level into a single variable needs to be done. For instance, it can be achieved by incorporating principal component analysis (i.e., dimension-reduction) on real-valued data. Lastly, there have been developed many niching methods that can preserve alternative candidates. Crowding, sharing, and spatial separation have been widely known. The concept of niching has given rise to *restricted tournament replacement* [89]. There is no difficulty in incorporating those niching techniques into rBOA and MrBOA.

To test the developed hrBOA and MhrBOA, moreover, research on designing real-valued hierarchical problems is imperative.

7.3 Concluding Remarks

The final goal of this book is to offer effective black-box optimization tools for solving a broad class of real-world problems quickly, accurately, and reliably by employing genetic and evolutionary computation (GEC). In this regard, five primary issues in GEC have been investigated. First, bridge the gap between theory and practice of GEAs; thereby providing practical design guidelines. Second, exhibit the practical use of the suggested design methodology by designing a GA-based routing algorithm. Third, devise simple but efficient optimization algorithms in the context of simple EDAs, which effectively and speedily solve memory- and time-constrained problems without incorporating any prior knowledge about the problems. Fourth, develop competent optimization algorithms from the standpoint of advanced EDAs. Finally, design competent multiobjective optimization algorithms within the framework of multiobjective EDAs.

It is hoped that the work will have a lasting influence on future research work on GEC as well as computational optimization.

References

1. Ahn, C. W., Ramakrishna, R. S., Kang, C. G., and Choi, I. C. (2001). Shortest path routing algorithm using hopfield neural network. *Electronics Letters*, 37(19), pages 1176–1178.
2. Ahn, C. W., Ramakrishna, R. S., and Kang, C. G. (2002). Efficient clustering-based routing protocol in mobile *ad-hoc* networks. In *Proceedings of the Vehicular Technology Conference (VTC'02)*, pages 1647–1651.
3. Ahn, C. W. and Ramakrishna, R. S. (2002). A genetic algorithm for shortest path routing problem and the sizing of populations. *IEEE Transactions on Evolutionary Computation*, 6(6), pages 566–579.
4. Ahn, C. W. and Ramakrishna, R. S. (2003). Elitism-based compact genetic algorithms. *IEEE Transactions on Evolutionary Computation*, 7(4), pages 367–385.
5. Ahn, C. W., Goldberg, D. E., and Ramkrishna, R. S. (2004). Multiple-deme parallel estimation of distribution algorithms: Basic framework and application. *International Conference on Parallel Processing and Applied Mathematics – PPAM 2003, Lecture Notes in Computer Science 3019*, pages 544–551.
6. Ahn, C. W., Goldberg, D. E., and Ramakrishna, R. S. (2004). Real-coded Bayesian optimization algorithm: Bringing the strength of BOA into the continuous world. *Genetic and Evolutionary Computation Conference – GECCO 2004, Lecture Notes in Computer Science 3102*, pages 840–851.
7. Ahn, C. W. and Ramakrishna, R. S. (2005). Building-Block supply in real-coded genetic algorithms: A first step on the population-sizing model. *IEICE Transactions on Fundamentals.* (in revision).
8. Ali, M. K. and Kamoun, F. (1993). Neural networks for shortest path computation and routing in computer networks. *IEEE Transactions on Neural Networks*, 4(6), pages 941–954.
9. Bäck, T. (1994). Selective pressure in evolutionary algorithms: A characterization of selection mechanisms. In *Proceedings of the First IEEE Conference on Evolutionary Computation*, pages 57–62.
10. Bäck, T. (1995). Generalized convergence models for tournament- and (μ, λ)-selection. In *Proceedings of the 6th International Conference on Genetic Algorithms*, San Francisco, CA, pages 2–8.

11. Bäck, T., Fogel, D. B., and Michalewicz, Z. (1997). *Handbook of Evolutionary Computation.* New York: Institution of Physics Publishing and Oxford University Press.

12. Baker, J. E. (1985). Adaptive selection methods for genetic algorithms. In *Proceedings of International Conference on Genetic Algorithms and Their Applications.* pages 101–111.

13. Baluja, S. (1994). Population-based incremental learning: A method for integrating genetic search based function optimization and competitive learning. *Technical Report CMU-CS-94-163,* Carnegie Mellon University.

14. Baluja, S. and Davies, S. (1997). Using optimal dependency-trees for combinatorial optimization: Learning the structure of the search space. In *Proceedings of the 14th International Conference on Machine Learning,* pages 30–38.

15. Baraglia, R., Hidalgo, J. I., and Perego, R. (2001). A hybrid heuristic for the traveling saleman problem. *IEEE Transactions on Evolutionary Computation,* 5(6), pages 613–622.

16. Booker, L. B. (1982). *Intelligent behavior as an adaptation to the task environment.* Doctoral dissertation, University of Michigan, Ann Arbor, MI.

17. Bosman, P. A. N. and Thierens, D. (2001). Advancing continuous IDEAs with mixture distributions and factorization selection metrics. In *Proceedings of OBUPM workshop at the Genectic and Evolutionary Computation Conference,* pages 208–212.

18. Bosman, P. A. N. and Thierens, D. (2002). Multiobjective optimization with diversity preserving mixture-based iterated density estimation evolutionary algorithm. *International Journal of Approximate Reasoning,* 31(3), pages 259–289.

19. Bosman, P. A. N. and Thierens, D. (2003). The balance between proximity and diversity in multiobjective evolutionary algorithm. *IEEE Transactions on Evolutionary Computation,* 7(2), pages 174–188.

20. Bosman, P. A. N. (2003). *Design and application of iterated density-estimation evolutionary algorithms.* Doctoral Dissertation, Utrecht University, TB Utrecht, The Netherlands.

21. Brindle, A. (1981). *Genetic algorithms for function optimization.* Doctoral dissertation, University of Alberta, Edmonton, Canada.

22. Cantú-Paz, E. (2000). *Efficient and accurate parallel genetic algorithms.* Boston, MA: Kluwer Academic Publishers.

23. Chellapilla, K. and Fogel, D. B. (2001). Evolving an expert checkers playing program without using human expertise. *IEEE Transactions on Evolutionary Computation,* 5(4), pages 422–428.

24. Chen, J. H. (2004). *Theory and Applications of Efficient Multiobjective Evolutionary Algorithms,* Doctoral dissertation, Feng Chia University, Taichung, Taiwan, R.O.C.

25. Chen, Y. P. (2004). *Extending the Scalability of Linkage Learning Genetic Algorithms: Theory and Practice,* Doctoral dissertation, University of Illinois at Urbana-Champaign, Urbana, IL.

26. De Bonet, J. S., Isbell, C. L., and Viola, P. (1997). MIMIC: Finding optima by estimating probability densities. *Advances in neural information processing systems,* 9, pages 424–431.

27. Deb, K. and Goldberg, D. E. (1993). Analyzing deception in trap functions. In *Foundations of Genetic Algorithms,* San Mateo, CA: Morgan Kaufmann, pages 93–108.

28. Deb, k. and Agarwal, S. (1995). Simulated binary crossover for continuous search space. *Complex Systems*, 9(2), pages 115–148.

29. Deb, k., Pratap, A., Agarwal, S., Meyarivan, T. (2002). A fast and elitist multiobjective genetic algorithm: NSGA-II. *IEEE Transactions on Evolutionary Computation*, 6(2), pages 182–197.

30. De Groot, M. H. (1970). *Optimal statistical decisions.* New York, McGraw-Hill.

31. De Jong, K. A. (1975). *An analysis of the behavior of a class of genetic adaptive systems*, Doctoral dissertation, University of Michigan, Ann Arbor, MI.

32. Dumitrescu, D., Lazzerini, B., Jain, L. C., and Dumitrescu, A. (2000). *Evolutionary computation*, Boca Raton, FL: CRC Press.

33. Etxeberria, R. and Larrañaga, P. (1999). Global optimization using Bayesian networks. In *Proceedings of Second Symposium on Artificial Intelligence (CIMAF-99)*, pages 332–339.

34. Fieldsend, J. E., Everson, R. M., and Singh, S. (2003). Using unconstrained elite archives for multiobjective optimization. *IEEE Transactions on Evolutionary Computation*, 7(3), pages 305–323.

35. Figueiredo, M. A. T. and Jain, A. K. (2002). Unsupervised learning of finite mixture models. *IEEE Transactions on Pattern Analysis and Machine Intelligence*, 24(3), pages 381-395.

36. Fonseca, C. M. and Fleming, P. J. (1995). An overview of evolutionary algorithms in multiobjective optimization. *Evolutionary Computation*, 3(1), pages 1–16.

37. Fonseca, C. M. and Fleming, P. J. (1998). Multiobjective optimization and multiple constraint handling with evolutionary algorithms – partI: a unified formulation. *IEEE Transactions on System, Man, and Cybernetics*, 28(1), pages 26–37.

38. Goldberg, D. E. (1989). *Genetic algorithms in search, optimization, and machine learning.* Reading, MA: Addison-Wesley.

39. Goldberg, D. E. and Rudnick, M. (1991). Genetic algorithms and the variance of fitness. *Complex Systems*, 5(3), pages 265–278.

40. Goldberg, D. E., Deb, K., and Clark, J. H. (1992). Genetic algorithms, noise, and the sizing of populations. *Complex Systems*, 6(4), pages 333-362.

41. Goldberg, D. E. (2002). *The design of innovation: Lessons from and for competent genetic algorithms.* Kluwer Academic Publishers.

42. González, C., Lozano, J. A., and Larrañaga, P. (2002). Mathematical modelling of MUDAc algorithm with tournament selection. Behavior on linear and quadratic functions. *International Journal of Approximate Reasoning*, 31(3), pages 313–340.

43. Han, K. H. and Kim, J. H. (2002). Quantum-Inspired Evolutionary Algorithm for a Class of Combinatorial Optimization. *IEEE Transactions on Evolutionary Computation*, 6(6), pages 580–593.

44. Harik, G. (1999). Linkage learning via probabilistic modeling in the ECGA. *IlliGAL Technical Report No. 99010*, University of Illinois at Urbana-Champaign, Illinois Genetic Algorithms Laboratory, Urbana, IL.

45. Harik, G., Cantú-Paz, E., Goldberg, D. E., and Miller, B. L. (1999). The gambler's ruin problem, genetic algorithms, and the sizing of populations. *Evolutionary Computation*, 7(3), pages 231-253.

46. Harik, G., Lobo, F. G., and Goldberg, D. E. (1999). The compact genetic algorithm. *IEEE Transactions on Evolutionary Computation*, 3(4), pages 287–297.

47. Hartigan, J. (1975). *Clustering algorithms*, New York, John Wiley & Sons.
48. Haupt, R. L. and Haupt, S. E. (1998). *Practical genetic algorithms*. New York, John Wiley & Sons.
49. He, J. and Yao, X. (2002). From an individual to a population: An analysis of the fist hitting time of population-based evolutionary algorithms. *IEEE Transactions on Evolutionary Computation.* 6(5), pages 495–511.
50. Heckerman, D., Geiger, D., and Chickering, D. M. (1994). Learning Bayesian networks: The combination of knowledge and statistical data. *Technical Report MSR-TR-94-09*, Redmond, WA: Microsoft Research.
51. Henrion, M. (1988). Propagating uncertainty in Bayesian networks by probabilistic logic sampling. *Uncertainty in Artificial Intelligence 2*, pages 149–163, Amsterdam.
52. Hidalgo, J. I., Lanchars, J., Ibarra, A., and Hermida, R. (2002). A hybrid evolutionary algorithm for multi-FPGA systems design. In *Proceedings of Euromicro Symposium of Digital System Design (DSD 2002)*, pages 60–67.
53. Holland, J. H. (1975). *Adaptation in natural and artificial system*. Ann Arbor, MI: University of Michigan Press.
54. Horn, j., Nafpliotis, N., and Goldberg, D. E. (1994). A niched pareto genetic algorithm for multiobjective optimization. In *Proceedings of the IEEE Conference on Evolutionary Computation (ICEC'94)*, pages 82–87.
55. Hoyweghen, C. V. (20010). Detecting spin-flip symmetry in optimization problems. In *Theoretical Aspects of Evolutionary Computing (Natural Computing Series)*, Kallel, L., Naudts, B., and Rogers, A., Eds, Berlin, Germany: Springer-Verlag, pages 423–437.
56. Hue, X. (1997). Genetic algorithms for optimization: Background and applications. Edinburgh Parallel Computing Centre, University of Edinburgh, Scotland, Ver 1.0.
57. Inagaki, J., Haseyama, M., and Kigajima, H. (1999). A genetic algorithm for determining multiple routes and its applications. In *Proceedings of the IEEE International Symposium on Circuits and Systems*, pages 137–140.
58. Jünger, M., [Online].
    http://www.informatik.uni-koeln.de/ls_juenger/projects/sgs.html.
59. Khan, N., Goldberg, D. E., and Pelikan, M. (2002). Multiobjective Bayesian optimization algorithms. *IlliGAL Report No. 2002009*, University of Illinois at Urbana-Champaign, Urbana, IL.
60. Khan, N. (2003). *Bayesian optimization algorithms for multiobjective and hierarchically difficult problems*. Master thesis, University of Illinois at Urbana-Champaign, Urbana, IL.
61. Larrañaga, P., Etxeberria, R., Lozano, J. A., and Peña, J. M. (2000a). Combinatorial optimization by learning and simulation of Bayesian networks. In *Proceedings of the 16th Conference on Uncertainty in Artificial Intelligence*, pages 343–352.
62. Larrañaga, P., Etxeberria, R., Lozano, J. A., and Peña, J. M. (2000b). Optimization in continuous domains by learning and simulation of Gaussina networks. In *Proceedings of the Genetic and Evolutionary Computation Conference Workshop Program*, pages 201–204.
63. Larrañaga, P., Lozano, J. A., and Bengoetxea, E. (2001). Estimation of distribution algorithms based on multivariate normal and Gaussian networks. *Technical Report KZZA-1K-1-01*, Department of Computer Science and Artificial Intelligence, University of the Basque Country.

64. Larrañaga, P. and Lozano, J. A. (2002). *Estimation of distribution algorithms: A new tool for evolutionary computation.* Kluwer Academic Publishers.

65. Lauritzen, S. L. (1996). *Graphical models.* Oxford, Clarendon Press.

66. Lee, C. Y. and Yao, X. (2001). Evolutionary algorithms with adaptive lévy mutations. In *Proceedings of the IEEE Congress on Evolutionary Computation (ICEC 2001)*, pages 568–575.

67. Leung, Y., Li, G., and Xu, Z. B. (1998). A genetic algorithm for the multiple destination routing problems. *IEEE Transactions on Evolutionary Computation*, 2(4), pages 150–161.

68. Lu, H. and Yen, G. G. (2003). Rank-density-based multiobjective genetic algorithm and benchmark test function study. *IEEE Transactions on Evolutionay Computation*, 7(4), pages 325–343.

69. Macready, W. G. and Wolpert, D. H. (1998). Bandit problems and the exploration/exploitation tradeoff. *IEEE Transactions on Evolutionary Computation*, 2(1), pages 2–22.

70. McLachlan, G. and Peel, D. (2000). *Finite mixture models.* New York, John Wiley & Sons.

71. Michalewicz, Z. (1992). *Genetic Algorithms + Data Structures = Evolution Programs.* Berlin, Heidelberg, New York: Springer-Verlag.

72. Moy, J. (1994). Open shortest path first protocol. *RFC 1583.*

73. Mühlenbein, H. and Schlierkamp-Voosen, D. (1993). Predictive models for the breeder genetic algorithm: I. Continuous paramenter optimization. *Evolutioanry Computation*, 1(1), pages 25–49.

74. Mühlenbein, H. and Schlierkamp-Voosen, D. (1993). The science of breeding and its application to the breeder genetic algorithm (BGA). *Evolutioanry Computation*, 1(4), pages 335–360.

75. Mühlenbein, H. and Paaß, G. (1996). From recombination for genes to the estimation of distributions I. Binary parameters. *Parallel Problem Solving from Nature – PPSN IV, Lecture Notes in Computer Science 1141*, pages 178–187.

76. Mühlenbein, H. and Mahnig, T. (1999). FDA – A scalable evolutionary algorithm for the optimization of additively decomposed function. *Evolutionary Computation*, 7(4), pages 353–376.

77. Müller, S. D., Marchetto, J., Airaghi, S., and Koumoutsakos, P. (2002). Optimization based on bacterial chemotaxis. *IEEE Transactions on Evolutionary Computation*, 6(1), pages 16–29.

78. Mostafa, M. E. and Eid, S. M. A. (2000). A genetic algorithm for joint optimization of capacity and flow assignment in packet switched networks. In *Proceedings of the 17th National Radio Science Conference*, pages C5-1–C5-6.

79. Munetomo, M., Takai, Y., and Sato, Y. (1998). A migration scheme for the genetic adaptive routing algorithm. In *Proceeding of the IEEE International Conference on Systems, Man, and Cybernetics*, pages 2774–2779.

80. Murthy, S. and Garcia-Luna-Aceves J. J. (1996). An efficient routing protocol for wireless networks. *ACM Mobile Networks Apllications Journal*, 1(2), pages 183–197.

81. Ocenasek, J. and Schwarz, J. (2002). Estimation of distribution algorithm for mixed continuous-discrete optimization problems. In *Proceedings of the 2nd International Symposium on Computational Intelligence*, pages 115–120.

82. Ocenasek, J. (2002). *Parallel estimation of distribution algorithms.* Doctoral dissertation, Brno University of Technology, Brno, Czech.

83. Ocenasek, J., Kern, S., Hansen, N., and Koumoutsakos, P. (2004). A mixed Bayesian optimization algorithm with variance adaptation. *Parallel Problem Solving from Nature – PPSN-VIII, Lecture Notes in Computer Sciences 3242*, pages 352–361.

84. Pan, H. and Wang, I. Y. (1991). The bandwidth allocation of ATM through genetic algorithm. In *Proceedings of the IEEE GLOBECOM'91*, pages 125–129.

85. Park, D. C. and Choi, S. E. (1998). A neural network based multi-destination routing algorithm for communication network. In *Proceedings of the IEEE International Joint Conference on Neural Networks*. pages 1673–1678.

86. Paul, T. K. and Iba, H. (2003). Reinforcement Learning Estimation of Distribution Algorithm. *Genetic and Evolutionary Computation Conference – GECCO 2003, Lecture Notes in Computer Science 2724*, pages 1259–1270.

87. Pelikan, M. and Mühlenbein, H. (1999). The bivariate marginal distribution algorithm. *Advances in Soft Computing – Engineering Design and Manufacturing*, pages 521–535.

88. Pelikan, M., Goldberg, D. E., and Cantú-Paz, E. (1999). BOA – The Bayesian optimization algorithm. In *Proceedings of the Genetic and Evolutionary Computation Computation Conference*, Morgan Kauffman, pages 525–532.

89. Pelikan, M. (2002). *Bayesian optimization algorithm: From single level to hierarchy*. Doctoral dissertation, University of Illinois at Urbana-Champaign, Urbana, IL.

90. Pelikan, M., Goldberg, D. E., and Lobo, F. G. (2002). A survey of optimization by building and using probabilistic models. *Computational Optimization and Applications*, 21(1), pages 5–20.

91. Pelikan, M., Sastry, K., and Goldberg, D. E. (2002). Scalability of the Bayesian optimization algorithm. *International Journal of Approximate Reasoning*, 31(3), pages 221–258.

92. Pelikan, M., Goldberg, D. E., and Tsutsu, S. (2003). Getting the best of both worlds: Discrete and continuous genetic and evolutionary algorithms in correct. *Information Sciences*, 156(3–4), pages 147–171.

93. Pelikan, M. and Goldberg, D. E. (2003). Hierarchical BOA solves Ising spin glasses and MAXSAT. *Genetic and Evolutionary Computation Conference – GECCO 2003, Lecture Notes in Computer Science 2724*, pages 1271–1282.

94. Reed, P. M., Minsker, B. S., and Goldberg, D. E. (2001). The practitioner's role in competent search and optimization using genetic algorithms. Presented at the *World Water and Environmental Resources Congress*, Washington, DC.

95. Rogers, A. and Bennett A. (1999). Genetic drift in genetic algorithm selection schemes. *IEEE Transactions on Evolutionary Computation*, 3(4), pages 298–303.

96. Rosenbrock, H. H. (1960). An automatic method for finding the greatest or least value of a function. *The Computer Journal*, 3(3), pages 175–184.

97. Rudlof, S. and Köppen, M. (1996). Stochastic hill climbing with learning by vectors of normal distributions. In *Proceedings of the First On-line Workshop on Soft Computing*, Nagoya, Japan.

98. Rudolph, G. (2001). Self-adaptive mutations may lead to premature convergence. *IEEE Transactions on Evolutionary Computation*, 5(4), pages 410–414.

99. Salomon, R. (1998). Evolutionary algorithms and gradient search: Simiarities and differences. *IEEE Transactions on Evolutionary Computation*, 2(2), pages 45–55.

100. Sastry, K. and Goldberg, D. E. (2000). On extended compact genetic algorithm. In *Proceedings of Late Breaking Papers in Genetic and Evolutionary Compuatation Conference*, San Francisco, CA, pages 352–359.

101. Sastry, K. and Goldberg, D. E. (2001). Modeling tournamnet selection with replacement using apparent added noise. In *Intelligent Engineering Systems Through Artificial Neural Networks*, New York: ASME, pages 129–134.

102. Sastry, K. (2001). *Evaluation-relaxation schemes for genetic and evolutionary algorithms*. Master's Thesis, University of Illinois at Urbana-Champaign, Illinois Genetic Algorithms Laboratory, Urbana, IL.

103. Schaffer, J. D. (1984). *Multiple objective optimization with vector evaluated genetic algorithms*. Doctroal Dissertation, Vanderbilt University, Nashville, TN.

104. Schaffer, J. D., Caruana, R. A., Eshelman, L. J., and Das, R. (1989). A study of control parameters affecting online performance of genetic algorithms for function optimization. In *Proceedings of the 3rd International Conference on Genetic Algorithms*, San Francisco, CA, pages 51–59.

105. Schewefel, H. (1981). *Numerical optimization of computer models*. Chichester, John Wiley & Sons.

106. Sebag, M. and Ducoulombier, A. (1998). Extending population-based incremental learning to continuous search spaces. *Parallel Problem Solving from Nature – PPSN V, Lecture Notes in Computer Science 1498*, pages 418–427.

107. Sinha, A. and Goldberg, D. E. (2001). Verification and extension of the theory of global-local hybrids. In *Proceedings of the Genetic and Evolutionary Computation Conference (GECCO 2001)*, pages 591–597.

108. Shimamoto, N., Hiramatus, A., and Yamasaki, K. (1993). A dynamic routing control based on a genetic algorithm. In *Proceedings of the IEEE International Conference on Neural Networks*, pages 1123–1128.

109. Srinivas, N. and Deb, K. (1995). Multiobjective optimization using nondominated sorting in genetic algorithms. *Evolutionary Computation*, 2(3): pages 221–248.

110. Stalling, W. (1998), *High-speed networks: TCP/IP and ATM design principles*. Englewood Cliffs, NJ: Prentice-Hall.

111. Syswerda, G. (1989). Uniform crossover in genetic algorithms. In *Proceedings of the 3rd International Conference on Genetic Algorithms*, San Mateo, CA: Morgan Kaufmann, pages 2–9.

112. Thierens, D. (1997). Selection schemes, elitist recombination, and selection intensity. In *Proceedings of the 7th International Conference on Genetic Algorithms*, San Francisco, CA, pages 152–159.

113. Törn, A. and Žilinskas, A. (1989). Global optimization. *Lecture Notes in Computer Science 350*. Springer–Verlag, Berlin.

114. Tsutsui, S., Pelikan, M., and Goldberg, D. E. (2001). Evolutionary algorithm using marginal histogram models in continuous domain. *IlliGAL Technical Report No. 2001019*, University of Illinois at Urbana-Champaign, Illinois Genetic Algorithms Laboratory, Urbana, IL.

115. Tsutsui, S. (2002). Probabilistic model-building genetic algorithms in permutation representation domain using edge histogram. *Parallel Problem Solving from Nature – PPSN VII, Lecture Notes in Computer Science 2439*, pages 224–233.

116. Tufte, G. and Haddow, P. C. (1999). Prototyping a GA pipeline for complete hardware evolution. In *Proceedings the 1st NASA/DoD Workshop on Evolvable Hardware*, pages 76–84.

117. Van Veldhuizen, D. A. (1999). *Multiobjective evolutionary algorithms: Classification, analysis, and new innovations*, Doctoral dissertation, Graduate School of Engineering of the Air Force Institute of Technology, WPAFB, Ohio.

118. Xiawei, Z., Changjia, C., and Cang, Z. (2000). A genetic algorithm for multicasting routing problem. In *Proceedings of the Internation Conference on Communication Technology (WCC-ICCT 2000)*, pages1248–1253.

119. Yao, X., Liu, Y., and Lin, G. (1999). Evolutionary programming made faster. *IEEE Transactions on Evolutionary Computation*, 3(2), pages 82–102.

120. Zhang, Q. and Leung, Y. W. (1999). An orthogonal genetic algorithm for multimedia multicast routing. *IEEE Transaction on Evolutionary Computation*, 3(1), pages 53–62.

121. Zhang, Q. and Mühlenbein, H. (2004). On the convergence of a class of estimation of distribution algorithms. *IEEE Transactions on Evolutionary Computation*, 8(2), pages 127–136.

122. Zitzler, E. (1999). *Evolutionary algorithms for multiobjective optimization: Methods and applications*. Doctoral dissertation, Swiss Federal Institute of Technology (ETH), Zürich, Switzerland.

123. Zitzler, E. and Thiele, L. (1999). Multiobjective evolutionary algorithms: A comparative case study and the strength Pareto approach. *IEEE Transactions on Evolutionary Computation*, 3(4), pages 257–271.

124. Zitzler, E., Laumanns, M., and Thiele, L. (2002). SPEA2: Improving the strength Pareto evolutionary algorithm. In *Proceedings of Evolutionary Methods for Design, Optimization, and Control*, pages 95–100.

Index